Transforming American Sc

Transforming American Science documents the ways in which federal funds catalyzed or accelerated changes in both university culture and the broader system of American higher education during the post-World War II decades.

The events of the book lie within the context of the Cold War, when pressure to maintain parity with the Soviet Union impelled more generous government spending and a willingness of some universities to reorient their missions in the service of country and of science. The book draws upon a substantial amount of archival research conducted in various university archives (MIT, Berkeley, Stanford) as well as at the Library of Congress, the National Archives, and various presidential libraries. Author Jonathan Engel considers the repurposing of the wartime Manhattan Engineering District and the Office of Naval Research to robust peacetime roles in supporting the nation's expanding research efforts, along with the birth of the National Science Foundation, space exploration, and atoms for peace among other topics.

This volume is the perfect resource for all those interested in Cold War history and in the history of American science and technology policy.

Jonathan Engel is Professor of Public Affairs at the Marxe School of Public and International Affairs at Baruch College (CUNY). He writes about the historical evolution of US health and social welfare policy. His previous books include *Doctors and Reformers: Discussion and Debate on Health Policy, 1925–1950*; *Poor People's Medicine: Medicaid and American Charity Care Since 1965*; and *Fat Nation: Obesity in America Since 1945*.

Routledge Studies in the History of Science, Technology and Medicine

For more information about this series, please visit: www.routledge.com/Routledge-Studies-in-the-History-of-Science-Technology-and-Medicine/book-series/HISTSCI

Transforming American Science

Universities, the Government, and the Cold War

Jonathan Engel

Routledge
Taylor & Francis Group

NEW YORK AND LONDON

First published 2023
by Routledge
605 Third Avenue, New York, NY 10158

and by Routledge
4 Park Square, Milton Park, Abingdon, Oxon, OX14 4RN

Routledge is an imprint of the Taylor & Francis Group, an informa business

© 2023 Jonathan Engel

ISBN: 978-1-032-42704-1 (hbk)
ISBN: 978-1-032-42705-8 (pbk)
ISBN: 978-1-003-36389-7 (ebk)

DOI: 10.4324/9781003363897

Typeset in Bembo
by Apex CoVantage, LLC

Contents

Figures

Acknowledgments

Research for this book was supported by a grant from the Dean's Research Fund at the Marxe School of Public and International Affairs.

My work for this volume benefited from input and suggestions of colleagues and friends, including Carla Robbins, Nancy Aries, Jerry Mitchell, Jack Krauskopf, Lewis Friedman, and James Hershberg.

My research was abetted by friends who hosted me on my travels. Thank you to Samuel Engel and Anne Freeh Engel, Andrew Engel and Karen Barr, and Joel Ginsberg.

Many thanks to my wife, Rozlyn Engel, for her thoughtful feedback and for suggesting the topic.

1 Introduction

The Government's Role

In the two years after World War II ended, the number of American men and women in uniform dropped from 12 million to 1.8 million. Large numbers of naval vessels, planes, tanks, and trucks were decommissioned, mothballed, or sold off. Hundreds of thousands of weapons, along with millions of rounds of ammunition, were placed in long-term storage, destroyed, or sold to friendly nations. Uniforms, canteens, boots, helmets, and mess kits were sold in bulk to army surplus outfits to pass on to eager consumers of army and navy surplus. The nation drew down rapidly as its servicemen and women found civilian employment and busied themselves with the tasks of buying homes, raising children, and accumulating wealth. The nation had never in its history maintained a large, standing, peacetime army, and political leaders deemed this to be desirable. In 1939, the US Army had ranked nineteenth in size in the world – just behind that of Portugal – and more than a few members of Congress thought that this was a reasonable size to return to.

There were exceptions, however. The three wartime locations of the Manhattan Engineering District were maintained and turned over to the newly created Atomic Energy Commission (AEC) in 1946. With 50,000 full-time employees, five major installations, and landholdings equal to half the state of New Jersey, the AEC instantly became a substantial federal agency. At the same time military research, centered in the Naval Research Laboratory in Washington and in various Army ballistics laboratories, was both retained and enlarged.

Government investment in all science, in fact, rose sharply after the war. The substantial wartime buildup in arms development, which had produced the atomic bomb, radar, and the proximity fuse, now became a permanent fixture of the American government. Guided by the vision of Vannevar Bush, the MIT professor of electrical engineering who had become the government's wartime administrator of research and development, the nation committed itself to permanently funding both basic and applied research in government labs through a growing network of grants and contracts with the nation's leading research universities. Between 1950 and 1961, government funding for research and development steadily grew 14 percent annually, and

DOI: 10.4324/9781003363897-1

thereupon it started an even sharper upward trend in the tremendous effort to match Soviet accomplishments in space.[1] Government research funding *doubled* nearly every five years between 1945 and 1969.[2] Funding on research and development from the Department of Defense (DoD) grew from just over $300 million at the end of the war to $1.8 billion within a decade. Large wartime aggregations of scientists and engineers, notably at the Radiation Laboratory at MIT and at the Applied Physics Laboratory at Johns Hopkins, were reorganized, moved to new quarters, and seeded with ever more generous allocations. The Office of Naval Research (ONR) under the guidance of its research director, Alan Waterman, quickly became the nation's largest funder of both basic and applied research while the AEC expanded the wartime Manhattan Engineering District facilities to create several "national" laboratories, each managed and staffed by an affiliated university. By 1949, the AEC was supporting 800 doctoral fellows and the ONR was supporting 1,200. Together, the two agencies funded 96 percent of research in engineering and the physical sciences on university campuses.[3]

For the dozen American universities which most aggressively pursued research, the flow of government funds changed their institutional priorities and reshaped their cultures. Immediately after World War II, government grants and contracts constituted 5 percent of the income for the entire university sector, but by the end of the 1960s it had risen to 26 percent.[4] This understated the substantial influence of government funding on the research powerhouses – MIT, the California Institute of Technology, the University of California, Johns Hopkins, Stanford, Harvard, the University of Chicago, Columbia, and a half dozen of their peers – where government funding might constitute almost 50 percent of their academic budgets and where professors might buy their way out of their teaching duties entirely.

All of this was by design. In *Science, the Endless Frontier*, a short but influential 1945 report to President Truman, Vannevar Bush had described the critical role which science should play in postwar American economic life and military planning. American parity with the Soviet Union would depend not on raw manpower but on sophisticated armaments. Economic growth and technological progress would rest on basic scientific innovations produced in university labs but funded by the federal government. New and increasingly sophisticated defense systems would come from a broad array of university and industrial contractors supported by a large and varied program of grants and contracts from the Navy, the Air Force, the AEC, and the National Science Foundation (NSF). The need to "re-capitalize" America's store of basic scientific knowledge – depleted by wartime demand and falling university enrollment – could be facilitated only by government largesse. Only public funds could meet the "rapidly increasing demands of industry and the government for new scientific knowledge," wrote Bush.[5]

The result was an elevation of science and scientists in the public's esteem. Members of Congress could confidently raise budget allocations annually without fear of constituent backlash. Contracts and grants to universities and private industry allowed the federal agencies to pay scientists at rates substantially above the government pay scale while harnessing the energy and insights of academic scientists who were now freed from bureaucratic constraints. Thomas Gates, President Eisenhower's last Secretary of Defense, would write with some amazement: "All of a sudden, the scientists became very important. . . . They had great veto power. They became very important people. You paid a lot of attention."[6]

The combination of independent university investigators, industrial research firms, and large national laboratories led to numerous victories for American science and engineering. Atomic bombs, guided missiles, antimissile defenses, sonar, increasingly sophisticated radar, the hydrogen "Super" bomb, satellites, and manned space missions all signaled either explicit military superiority over the Soviet Union or implied technological and intellectual might. Military leaders expressed satisfaction with the investment; even the most advanced weapons were based on relatively small research budgets (as a portion of the general military budget) and the nation seemed to have a strong appetite for new aircraft, missiles, rockets, and nuclear weapons. A congressional committee on research observed in 1963 that "the federal government's marriage to research and development has been marked by an amazingly long and luxurious honeymoon."[7]

Funding science through government grants and contracts as well as internally in government labs made sense; only government had the financial wherewithal to fund a national effort on the scale that Bush had envisioned. Even the wealthiest American universities and foundations could draw on only modest endowment funds to finance faculty research, and most universities had no endowments at all. By contrast, the budget of just one government facility in 1945, Oak Ridge, exceeded the total investment of all universities in all faculty research combined. Simply put, only the federal government had the money to impel American science forward.

But funding science through government brought risks. The public preferred science with clear applications: more engineering than pure science. One field was not more noble than the other, and both contributed to human progress. They had different professional aims and could work either symbiotically or independently. "The basic difference between science and engineering," wrote IBM's director of research, Emanuel Piore, "is that science tries to understand nature and engineering tries to control nature."[8] The public generally preferred funding the latter.

But a preference for engineering research ("applied science" in common parlance) brought the risk of depleting the stock of fundamental scientific insights. While engineering did not always require science, it generally did. The great wartime successes of fission and radar rested on fundamental work in nuclear physics from the 1920s and on electrochemistry from before. Perfecting microwave radar was an engineering challenge, but discovering the existence of radio waves had been scientific. Government tended to favor the former for its usefulness.[9] Politicians voting on budgets relied on voters who looked for government to improve their lives. Winning wars counted for a lot; understanding the universe, less so.

Applied science had a tendency to drive out basic science through its popular appeal, but the nature of applied science was also simply inhospitable to fundamental science. Big applied science – the type of huge wartime development programs employing many thousands of researchers and engineers – was hierarchical. The Manhattan Engineering District which had produced the atom bomb had been a branch of the Army. Its brilliant senior physicist, J. Robert Oppenheimer, had exercised great authority but ultimately ceded administrative oversight to army directive. But cumbersome hierarchies tended to work toward established goals while marginalizing scientific creativity. Merle Tuve, the director of the large industrial laboratory which produced the proximity fuse during the war, reflected some years later:

> Huge new synchrotrons and cosmotrons and electronic computers, and polar expeditions and balloon and rocket flights and great government laboratories costing more each year than the total academic cost of many of our greatest universities – all of these conspicuous aspects of our new national devotion to science are subsidiary and peripheral. They do not serve appreciably to produce or develop creative thinkers and productive investigators. At best they serve in a brief or a rather incidental way, and at worst they devour.[10]

Tuve's insight can be seen running through the decisions covered in this book. The challenge for the architects of postwar American science was not funding applied work; *that* could be ably done through the military and later through the AEC, with its roots in wartime nuclear weaponry. Rather, the challenge lay in funding basic work. Basic science was meandering and undirected – its very nature was whimsical. It explored the unknown with no clear expected outcome. Basic scientists not only did not thrive amid hierarchy; but they temperamentally rejected it. Where applied military research was directed and secretive, basic science was eccentric and open. Scientists were collaborative by nature, sharing their data, insights, and thinking with other scientists. And science was universal. A German physicist might pair with a Russian mathematician and a Japanese theorist. Even at the height of the Cold War, American physicists made efforts to stay in touch with their Soviet counterparts. Jingoism was simply inconsistent with the endeavor.

Above all, scientists valued freedom: freedom to organize their time, to focus on questions of interest to them, to design a research program, and to share their ideas. While not wholly antagonistic to accountability, scientists tended to understand each other best, relying on a system of sharply critical peer reviews to excise sloppy thinking and on informal interactions to hone ideas and sharpen insights. Leonard Loeb, a physicist at Berkeley, described his fellow scientists as wedded to a tradition of "precision of definition, skepticism, and criticism" without which science could not advance. "Authoritarianism is repugnant to them," he wrote. "They speak a common language arrived at by free exchange of ideas and mutual agreement."[11]

The drive toward freedom tended to push pure scientists out of government labs and into universities. The academic culture harbored in the nation's research universities was distinct from that found in government shops: less accountable, more innovative, and with fewer parameters and looser expectations. Academic appointments did not come freighted with an expectation of deliverables – the products and processes demanded by contractors in salaried work. Even the most elite of government labs, such as those found at the Naval Research Laboratory, tended to work toward a specific product or series of products. Universities could tolerate a more haphazard approach to productivity in ways which civil service and military managers simply could not. "Public employment has become, in a very real sense, a hazardous occupation," explained David Lilienthal, the chairman of the AEC to an assembled group of scientific leaders in 1948. "The growing concern of scientific and technical and managerial people is evident on every hand. The trend shows that as between private industry, educational institutions, and government, government is regarded as the least desirable employer to most scientists."[12]

In an effort to maintain the productive relationships which had been so critical in winning World War II, the military renewed its commitment to university contractors at the war's end including MIT, the California Institute of Technology, Johns Hopkins and, increasingly, Stanford. Stanford and MIT then extended these relationships to private subcontractors operating in the areas around the schools, creating the nation's original high-technology regions in Silicon Valley and Boston's Route 128. Each additional step of remove from government employment seemed to invite greater freedom of scientific inquiry, which in turn proved beneficial to innovation on both the basic and the applied fronts. Entire laboratories were funded through the military, the AEC, and later National Aeronautics and Space Agency (NASA) and turned over for operation and direction to university contractors. First came the AEC's national laboratories which included Los Alamos, Argonne, Oak Ridge, and Brookhaven; these were followed by the Air Force's Lincoln Lab, NASA's Jet Propulsion Lab, and the Navy's Applied Physics Lab. Notably, each of these labs was partnered with a university manager, who staffed and operated the facility, creating a uniquely hybrid construction – the government-owned university-managed laboratory. Even the most focused of weapons development programs seemed to benefit from the looser administrative constraints found at these hybrid government–university labs, where academics held reign and doctoral and postdoctoral students conducted the day-to-day work. In those labs where the military maintained too tight a hold, science tended to suffer. Burton Klein, fellow at RAND, wrote in 1958, "Research and Development is being crippled by the official refusal to recognize that technological progress is highly unpredictable, by the delusion that we can advance rapidly and economically by planning the future in detail."[13]

Whether in basic or applied fields, military need drove the agenda in the physical sciences in the quarter century after World War II. Advances in arms and armor had always played an outsized role in military supremacy, whether the long bows at Agincourt or the iron-cladding of World War I battleships.[14] Many military historians identified the decisive role played by the armored tank in finally ending the trench-bound stalemate of World War I. And during World War II, combat strategies seemed to pivot from one decisive technology to the next, which included submarines, sonar, radar, proximity fuses, precision bombing, and the atomic bomb.

In the 25 years after World War II, military funding dominated the physical sciences in the United States. In 1951, the inaugural year of the NSF, NSF funding for scientific research totaled $1.1 million while total funding from the DoD was $178 million and from the AEC $121 million. Even in the realm of basic science, the stated bailiwick of the NSF, funding from the DoD dwarfed funding from the NSF by 20 times.[15] And while the NSF budget would grow over time to $8 million by 1954, and to $140 million by 1964, it would always trail military research funding by a wide margin. By 1964, funding from the three main branches of the armed forces along with the Advanced Research Projects Agency (ARPA) totaled just over $7 billion, of which just about a tenth ($700 million) was designated to support basic research. When the tremendous amount of contractual research coming from NASA was added to the total, government funding of the engineering and physical sciences nearly doubled again, to just over $15 billion. The NSF, in theory the nation's preeminent champion of basic science, funded just under 1 percent of that.[16]

Eventually, the NSF would play a more significant role. The NASA budget declined rapidly after 1966, and the NSF budget grew nearly 20 percent annually through the remainder of the 1960s, even as the DoD began to pull back on its commitment to basic science. By 1970, the NSF could claim to fund nearly half of the nation's fundamental work in the physical sciences, and at smaller, less research-intensive universities the NSF grants actually rose in importance given the uneven dispersion of defense research funds.[17]

Military research tended to stay military, just as NASA research tended to remain in space. The unique needs of missile guidance, naval vessel propulsion, nuclear war, and space travel dictated scientific work which produced few civilian benefits. Weather and communications satellites and civilian nuclear power plants were notable exceptions to this rule, but most other efforts to apply military research to the civilian sector failed. An ill-fated effort to produce a nuclear-powered maritime freighter, for example, resulted in the *Savannah*, which lost millions of dollars in its decade at sea. A series of bizarre proposals in the 1950s to use atomic bombs to excavate artificial harbors or to enlarge inlets went nowhere. The one effort at civilian supersonic air transport, the Concord, produced annual losses for its parent airlines before being permanently grounded in 2003.[18] And with the noted exceptions of advances in certain fields of material engineering, the Apollo moon flights produced little of use for non-astronauts. This was not to say that military research was for naught, only that applied research was just that – applied. And military applications tended to be highly tailored to the extreme and unusual conditions of combat.

An important effect of federal science policy during these decades was its tendency to concentrate talent and solidify the leadership of the nation's most elite schools. In 1950, over half of all federal science funding went to just five universities, and while that number grew over time, by 1970 over 80 percent of the funding still went to just 20 schools. Competing for federal grants and research contracts was open only to the most well-resourced universities with the strongest faculty and the best laboratories. At the same time, the funds themselves worked to further concentrate research in the schools which were already strongest at the end of World War II. The military and the AEC sent their funds where they had produced past successes. Those schools, in turn, were able to use funding to lure the most promising young researchers to work with well-funded senior scientists. The best post-doctoral fellows and PhD candidates followed faculty talent, so that within a decade a mere dozen universities could draw virtually all of the best young scientists and graduate students to their doors for training, coauthorship, mentoring, and access to labs and scientific equipment. These same universities became primary contractors with the AEC to run the national labs or else developed partnerships with the Air Force, Navy, and NASA to staff and manage labs dedicated exclusively to those agencies.

A few new schools were able to join the ranks of the elite during this time. Rochester and Pennsylvania were able to partner with a consortium of universities to run the Brookhaven National Laboratory, and several campuses of the University of California were able to join Berkeley at the front ranks of biomedical and clinical research. Most notably, Stanford University pulled itself into the highest tier of research universities, aggressively recruiting top-notch faculty to its engineering programs and de-emphasizing teaching in favor of research. In 1966, the AEC built the world's largest linear accelerator on the Stanford campus which further cemented the university's reputation as the leading institution on the West Coast for work in engineering and the physical sciences.

Another factor which accentuated concentration of talent was the rise of high-technology clusters around the most research-intensive universities. Entrepreneurs established technology companies near Stanford and MIT to draw on faculty expertise and to hire recent graduates. Once established, these new firms often subcontracted with universities on government contracts and grants. The presence of these firms made their neighboring universities yet more attractive places to work and to study, given the possibility of outside consulting fees for faculty and of internships and employment for students. Soon, universities around the country made efforts to replicate the model, with whole cities and regions trying to recreate the success of Silicon Valley and Route 128. New Haven's Science Park, Raleigh's Research Triangle, Albany's nanotechnology campus, Madison's Madworks accelerator, and New York's Roosevelt Island incubator were all efforts to replicate the tremendously successful model pioneered by MIT and Stanford.

This flow of funds from the military and the AEC supercharged research on select campuses but at a cost. Faculty who brought their own contracts and grants with them became less beholden to their host universities as they turned their focus to professional peers who would review their grants applications, manuscripts, and promotion packages. Attention to teaching declined as well-funded faculty used their grants and contracts to buy down their teaching obligations and focus on research and on mentoring graduate students. Faculty hired at research universities understood that promotion, tenure, and career advancement would come not from winning approval of colleagues and supervisors from within their universities but rather from gaining endorsements from their professional peers outside of their universities. Eventually, many faculty came to view themselves as working *at* a university rather than *for* a university. External funding slowly eroded faculty commitment to university governance and priorities. The NSF recognized the phenomenon and identified it as a "flight from teaching."[19]

The flow of funds to the nation's research universities also distorted the sorts of questions on which scientists in the physical and engineering sciences focused. Scientists needed to earn their salaries, and with hiring and promotion focused on gaining outside funding, scientists understandably biased their research to those questions and fields of greatest interest to the military and the AEC. Whole departments were established and faculty lines funded with an eye toward winning outside contracts. MIT's agreement to establish the Lincoln Laboratory and Caltech's willingness to partner with the Jet Propulsion Laboratory demonstrated university acquiescence in building faculty and graduate programs around military matters. The NSF reported that whole programs had been "patterned rather closely on the mission requirements of individual Federal agencies," principally the DoD, the AEC, and later, NASA.[20]

As the competition for research funds heated up, universities actually found themselves competing not against other universities but rather against commercial and industrial laboratories which might prove equally competitive in garnering military contracts. This was not necessarily bad; nonuniversities were capable of conducting research at the highest levels. However, it did suggest further erosion of the unique mission of the university which had historically centered on producing and disseminating knowledge in the service of truth. Universities prided themselves on their commitment to academic freedom and inquiry. Tenure, itself, had been conceived as a means of insulating faculty from the whims of the market and of public opinion. But as the search for outside funding intensified, tenure guarantees declined. Employment predicated on securing outside funding was hardly secure, and it certainly no longer shielded the academic scientist from the vagaries of the market. For the handful of super universities, research had become not merely an adjunct to teaching in the university mission but rather the mission itself, and thus the entire model of ivory tower sanctuary was undermined.

Concentrated talent was good for science but possibly not good for the nation. As the nation's top two-dozen universities pulled progressively far-ther away from the majority of American colleges and universities, the reach of their reputations and networks (their "brands") became more valuable. For those individuals with the good fortune to gain affiliation to the elite institutions, the brands were a boon to accessing elite career opportuni-ties, doctoral programs, postdoctoral fellowships, internships, clerkships, and apprenticeships. While it was true that ambitious doctoral and professional students had long relocated to receive the best possible training, now under-graduates too traveled farther and paid more to gain access to elite networks. The premium for attending an elite university grew concomitantly with the reputation of the faculty, despite the fact that many faculty members had only a modest commitment to teaching. Government funding brought wealth to select campuses – in talent, resources, infrastructure, and networks – while the non-elite American universities fell increasingly behind.

In understanding these trends, it is important to view them within the Cold War context in which they transpired. American fear of Soviet military supremacy in the three decades after World War II was intense and justified. Although information coming out of the USSR was imperfect, American analysts and scientists understood that the Soviet Union was investing enor-mous resources in maintaining scientific and military parity with the United States during these decades. While initial estimates at the end of World War II suggested that the Soviet Union would possess an atomic bomb by 1955, the Soviets actually exploded one midway through 1949. When the Americans committed themselves to building a fusion (hydrogen) bomb, that commitment flowed in large part from fear that the Soviets would get a hydrogen bomb first. In fact, the Soviets followed the first American fusion blast in 1954 with their own hydrogen bomb only one year later.

Through these decades, the Soviets devoted extraordinary national resources toward training scientists, mathematicians, and engineers, and their investments produced dividends. Despite per capita earnings only one-fourth that of the United States in 1939, the Soviets out-trained and out-educated the Americans in the physical and engineering sciences in the 1940s and 1950s, managing along the way to pull ahead in rocket technol-ogy and jet propulsion and drawing even in fission-based weaponry, nuclear naval propulsion, radar, sonar, and guidance systems. With the launch of Sputnik in 1957, the Soviets achieved superiority in space travel, which they cemented by putting the first person into space. And while the Americans would beat the Soviets in landing a man on the moon in 1969, that had only been accomplished through prodigious investment of national resources. At the height of Apollo, in 1966, NASA was consuming 4 percent of the federal budget and 1 percent of the entire GDP. The total investment in NASA through the 1960s, about $25 billion, was nearly tenfold greater than the investment in the Manhattan Project at the height of World War II.[21]

During the period that this book covers, scientific prowess became not simply a means to military supremacy but an aim in and of itself within the broader Cold War. The two superpowers sparred on consumer products, automotive technology, international scientific acclaim and, of course, space. Scientific prowess became a proxy for national strength, similar to the way in which Olympic athletes became surrogate soldiers in a war for international influence. The United States obsessed over falling behind the Soviets in the quality and quantity of its scientists and engineers, even after the nation could not employ all of the young scientists that it trained.

The place of science in American life shifted substantially during these decades. While historically science had been the effete pursuit of privileged aristocrats and intellectuals, it now became central to economic growth, military parity, and national honor. Congressional leaders hired staff members with scientific expertise to guide them in their legislative efforts, and the White House over several administrations created a complex web of science advisors, coordinating committees, and special assistants. Charles Falk, the director of planning for the NSF, noted in 1967 the speed of the transformation, writing, "It has only been a short time – really not until after World War II – that we actually recognized science as a major national resource."[22]

This is the story of the ways in which American science, and alongside it American scientific institutions, evolved over the quarter century after World War II, during the heyday of the Cold War. The decision to remain on a wartime footing within the realm of military technology and armaments, along with the unique demands of nuclear engineering and particle physics, produced a new hybrid structure of government-funded university-based science. The relationship, built on a complex arrangement of university labs, university-controlled national labs, and affiliated high-technology firms working closely with university-based scientists, proved extraordinarily fruitful.[23] By 1970, the United States led the world in the rankings of its research universities, in its publications in leading scientific journals, and in its international scientific garlands. It had been first on the moon, first with a substantially nuclearized naval fleet, and first with submarine-launched intercontinental ballistic missiles. And while the Soviets had been closely behind at every juncture (and at times ahead, as in the early days of the two nations' space programs), the Americans had largely won. We can still see the effects of this competition a half-century later. Recent rankings of the world's leading research universities have American schools holding over half of the top 20 spots; the Russians hold none.[24]

Key to this story is the unique relationship that evolved between the United States' leading research universities and the government during this time, as universities transformed themselves into research arms of the Navy, the Air Force, the AEC, and (to a lesser degree) NASA, along with the National Institutes of Health.[25] The unique arrangement of government-owned but university-controlled laboratories – a singular solution to the problems of overly rigid bureaucratic control over science – produced scientific breakthroughs in extraordinary quantity and quality. What had emerged, a university–government axis, continues to this day. The structure has been highly successful in producing world-class science but perhaps at a cost of the integrity of university culture. As the gap in fiscal and academic strength between United States' marquis research universities and its regional teaching colleges grows larger, we might ask ourselves if, in our pursuit of the best science, we have sacrificed our system of higher education. This book offers some insights into why we built those structures and relationships in the way that we did and into the benefits that they produced and costs that they incurred.

Notes

1 Bruce Smith, *American Science Policy Since World War II*, Brookings, 1990, pp. 38–39.
2 Director of the Office of Defense Research and Engineering to Chairman, Defense Science Board, December 31, 1968, CCF, 4: DoD Research Policy, p. 3.
3 Daniel Kevles, "K1S2: Korea, Science, and the State," in Peter Galison and Bruce Herly, eds., *Big Science: The Growth of Large Scale Research*, Stanford University Press, 1992, pp. 314–15.
4 Richard Lewontin, "The Cold War and the Transformation of the Academy," in David Montgomery, ed., *The Cold War and the University*, New Press, 1997, pp. 24–25.
5 Vannevar Bush, *Endless Horizons*, Public Affairs Press, 1946, p. 54.
6 Quoted in Charles Maier, " 'Introduction' to George Kistiakowsky," in *A Scientist at the White House*, Harvard University Press, 1976, p. xxix.
7 Defense Science Board, "Report on Department of Defense Research Policy," December 31, 1963, CCF, 4: DoD Research Policy, p. 2.
8 Piore to Seaborg, June 1, 1960, OSAST, 13: Basic Research.
9 See Bush, *Endless Horizons*, p. 55.
10 Merle Tuve, "Is Science Too Big for the Scientist?" *Saturday Review*, 42, June 6, 1959, p. 49.
11 Leonard Loeb, "Military Security in a Scientific Age," *Science*, 120, July 30, 1954, p. 157.
12 David Lilienthal, "Address to the American Association for the Advancement of Science," *IIR*, 17:2, September 16, 1948, p. 4.
13 Burton Klein, "A Radical Proposal for R and D," *Fortune*, May 1958, p. 112.
14 The French had roughly double the number of soldiers and bowmen at Agincourt as the English, but superior bowmanship won the day. Historians estimate French dead at 6,000, and English dead about one-tenth of that. Hot air balloons never did prove decisive (or even important), but Franklin's premonition was born out in World War II, where air cover became an essential component of modern warfare. The battle of Midway became the first naval engagement in history where the two fleets never came within firing range of each other.
15 Harold Orlans, *The National Science Foundation: A Review of the Foundation's Granting and Contracting for Research*, Brookings, 1965.
16 Memo from Director of Defense Research and Engineering to Chairman, Defense Science Board, December 31, 1968, CCF, 4: DoD Research Policy, pp. 5–6.

17 In 1970, Lincoln Lab and the Instrumentation ("I") Lab at MIT alone drew $100 million of military funding.

18 Technically, the Concorde routes for Air France and British Airways were profitable but only because the original cost of the planes was heavily subsidized. Had the purchase price of the aircraft reflected the true development costs, the routes would have failed to turn a profit over nearly 30 years.

19 National Science Board of the National Science Foundation, "A Public Policy for Graduate Education in the Sciences," 1968, LBJ, WHCF, FG 265, 297: National Science Board, p. 25.

20 Ibid., p. 24.

21 In constant dollars.

22 Charles Falk, "Science and Public Policy Activities in Universities," *Bulletin of the Atomic Scientists*, June 1968, p. 50.

23 See Lee DuBridge's analysis of GOGO, GOCO, and COCO labs during these years in "Science and National Security," *Science*, 120, December 1954, pp. 1081–85.

24 QS Rankings, Global Universities, 2020, www.topuniversities.com/university-rankings/world-university-rankings/2020.

25 This book largely does not focus on the parallel growth of university-based government-funded research in the biomedical sciences, although there are many parallels between the two stories.

2 Wartime Efforts

Early Efforts

In April 1945, a badly damaged B-29 airplane approached the recently captured island of Iwo Jima, returning from a bombing run over Japan. Low on fuel, and flying through a soupy atmosphere on its narrow approach path to the landing strip, the pilot relied on directions from a ground crew using a recently perfected radar gadget known as the Ground Controlled Approach (GCA) to land in what would otherwise be untenable conditions. "Today a B-29 still flies because of us, and eleven men feel we saved their lives," wrote a member of the GCA crew. Dozens of other planes would land in similar conditions over the following months, and their crews would later write letters to GCA operators to thank them. On one particularly bad day that Spring, GCA operators landed 20 planes safely. "The rest of us stood around and sighed with relief each time a new safe landing was made," remembered one operator.[1]

GCA was the product of the Radiation Laboratory ("Rad Lab") at the Massachusetts Institute of Technology (MIT). The Rad Lab, which grew out of the electrical engineering faculty in the 1920s, had spent much of the 1930s exploring and perfecting the use of radio signal echoing to identify and locate objects in the air and on the sea. In late 1941, a group of researchers hypothesized that microwave radar technology had become accurate enough to determine an aircraft's altitude, bearing, range, and speed to such a degree that a ground crew could conceivably guide a plane to a landing despite near-zero visibility from the cockpit. Nine months of intensive research produced a workable prototype, which was reliable enough by 1943 to be accepted by the Army Air Corps. The US Army and the US Navy as well as the Royal Air Force began to use the device in early 1944. In the last year of World War II, use of the device became commonplace in all theaters of the war. One crew member, writing from Verdun, wrote of a P-61 aircraft landing at night in the fog:

> They brought him around in a very tight pattern and this time he came in all right. Visibility was so bad he couldn't see more than one runway

DOI: 10.4324/9781003363897-2

light at a time. When he finally stopped he had twenty gallons of gas left – which for the 2,000 hp engines in a P–61 is just about one good cough.[2]

Radar was only one of several scientific and technological developments during World War II critical to allied victory. Long-range radio navigation, proximity fuses, anti-submarine detection (SONAR), penicillin, and, most famously, the atomic bomb each played a substantial role in wartime success.[3] Radar bombsites installed in allied planes allowed for precision bombing of the fog-shrouded Normandy beaches before the D-Day landings. Proximity fuses, which detonated artillery shells as they approached their targets, were instrumental in thwarting Japanese kamikaze attacks on US ships and in neutralizing German V-1 rocket attacks on England. SONAR wars erupted between the US and German navies as the technology was used to guide torpedoes toward their targets and then later to thwart those torpedoes and redirect them to decoys. And the atomic bomb accelerated the Japanese surrender and precluded the need for an American invasion of Japan.[4]

It was hardly inevitable that the US military would turn to scientific research to abet its war effort. For much of its first century, the federal government eschewed involvement with scientific pursuits. John Quincy Adams's plea for a national astronomical observatory was forcefully denied by Congress in 1807.[5] Congress *did* agree to establish the Smithsonian Institution in 1846, off the income of a bequest left by Englishman James Smithson but only after dallying through ten years of debate. When federal support for scientific research finally arrived, in the wake of the Civil War, it was largely focused on agricultural and industrial needs. In rapid succession, Congress established the Weather Bureau (1870); the Geological Survey (1878); the Office of Agricultural Experiment Stations (1887, hitched to funding for the Land Grant colleges); the Bureau of Soils (1901); the Bureau of Reclamation of Plant Industry (1902); the Forest Service (1906); the Bureau of Mines (1910); and the National Advisory Committee for Aeronautics (1915) – only the last having potential military application.

World War I pointed to the potential usefulness of scientific and technological research for the military, particularly as it pertained to naval ordnance and navigation. Woodrow Wilson recruited Thomas Edison to chair a scientific board to advise the Navy, although Edison saw little need for pure scientists rather than industrial engineers. He grudgingly recruited a single physicist, saying, "There ought to be one mathematical fellow around in case we need to calculate out something."[6] During the interwar years, all branches of the military took small steps toward integrating technological and ordnance research into their portfolios. The Navy established the Naval Research Laboratory in 1923, and the Army followed with the Signal Corps, the Chemical Corps, and the Ordnance Corps. Even so, the combined research budget of the Army and Navy totaled only $11 million in

1930, at a time when nonfederal (mostly industrial and agricultural) research expenditures were $143 million.

Other countries were less lackadaisical. Germany, in particular, invested heavily in weapons research through World War I in an effort to break the murderous stalemate of trench warfare that was devastating allied and axis troops alike. Germany achieved a substantial (though troubling) technological breakthrough in the form of chlorine gas attacks, first proposed by the Nobel laureate chemist Fritz Haber, who was already producing nitric acid for the German munitions industry. Haber's idea was to release gas from cylinders at a time when prevailing winds would float the gas over to allied trenches. The gas, being heavier than air, would descend into the trenches, asphyxiating enemy soldiers and forcing a retreat. In April 1915, German gas troop *pionierkommandos* near Ypres, Belgium opened valves on 5,730 cylinders of liquefied chlorine for over ten minutes, floating a five-foot wall of gas toward French Algerian troops to the west. Within a few minutes, the soldiers lay "choking and vomiting and dying" while those who could fled the trenches.[7]

Germany pioneered mustard gas in 1917, while the United States and its allies struggled to develop effective gas masks and countermeasures. By the end of the war, the warring countries had produced 22 different chemical agents which could be delivered in a variety of projectiles and bombs. Between one-third and one-half of all shells delivered in the final months of the war contained chemical armaments, which were collectively responsible for as many as 560,000 casualties.

Despite the contributions of scientific research to the war effort, however, in the immediate aftermath of World War I, the US government was uninterested in pursuing arms-related research. Through the 1920s, proponents of research funding such as Secretary of State Elihu Root and Secretary of Commerce Herbert Hoover tried in vain to convert the National Academy of Science into a vehicle for diverting funds toward government research projects, but as late as 1933, the National Resources Board blocked a request by the Science Advisory Board for $16 million to fund a proposed Recovery Program in Science Progress.[8] During those years, the bulk of US research efforts were carried on by industrial concerns. General Electric had established the General Electric Laboratory, in 1900 to develop new products, which was joined by industrial labs at Bell Telephone, Westinghouse and Dupont in the ensuing decades. By 1930, industrial research expenditures counted for 70 percent of the nation's $170 million research effort.[9]

University-based research, however, remained moribund. In the field of electrical engineering, for example, all researchers in the United States together published 442 technical articles between 1920 and 1925, of which only 54 (12%) had a university-based author. The nation's electrical engineering faculty were publishing an average of nine papers annually. In the specialized field of radio engineering, university-based researchers accounted for an even smaller portion of the total output: less than 5 percent. Electrical engineering professors at the majority of universities published nothing. Of the 54 university-based publications during that half-decade, 44 came out of just five schools, with over half being authored by just eight professors.[10] Frederick Terman, a professor (and later dean) of electrical engineering at Stanford wrote, "This makes one begin to suspect that perhaps most of our electrical engineering schools are trailers of the electrical industry in general, rather than leaders of progress."[11] The inevitable result was a new generation of engineers being trained with outdated techniques and information, producing poorly supervised thesis work on "antique" problems.[12]

The Rad Lab

In the late 19th century, Heinrich Hertz in Germany demonstrated the existence of radio waves. These electromagnetic waves, too long to be seen by the human eye, could carry information in their varying frequencies and amplitudes. Guglielmo Marconi successfully sent a message across the Atlantic Ocean using the waves in 1901. By 1920, commercial radio broadcasts used the waves to transmit news, music, and radio shows.

Radio waves could be used to locate and describe targets through bouncing a wave off of a target and detecting the echo wave. Even before World War I, armies in Europe and the United States worked to exploit this potential in an effort to "see" incoming ships and airplanes and also to help their own ships and planes to navigate. The great limitation on the technology was the length of the waves themselves. Long waves could travel fairly far through the atmosphere, but they tended to deliver a blurred picture of the target. They were hard to aim and had a tendency to bounce off of the surface of the ocean (when used to detect ships) thus distorting the signal. Using long-wave radar, operators could see the existence of a target, but they could not discern details such as size and shape. A cluster of three airplanes might appear as a single blip on a screen, for example, and the ships in a flotilla could be hard to distinguish.

The solution lay in using shorter radio waves to describe a target. These "short" or "micro" waves, still longer than those on the visible spectrum, could be better aimed when passing them through an aperture and could produce a much more detailed picture of a target. But producing microwaves posed substantial challenges to electrical engineers. One of the people most interested in the challenge was industrialist Alfred Loomis, who built a private laboratory devoted to applied electronics research near his home in Tuxedo Park, New York. Inviting outside researchers to work with him, he made progress on microwave radar through the 1930s and laid the groundwork for the Loomis Radio Navigation system (LRN, later LORAN), which would fundamentally change the manner in which ships and planes navigated.[13]

Many of the scientists working with Loomis through the 1930s were faculty members from MIT's department of electrical engineering, the nation's largest and foremost department in that discipline. Under the urging of the institute's ambitious president, Karl Compton, the department had increased its research output (it published nearly double the number of scientific articles through the 1920s as the nation's second most productive department – Johns Hopkins) and moved to tighten working relationships with industry researchers at General Electric, Raytheon, Philco, ITT, RCA, and elsewhere.

Like any elite academic department, electrical engineering at MIT could boast of highly intelligent faculty members and students. But broader policies at MIT facilitated its growing success through the 1930s. The institute had built up strong relationships with industry across disciplines through the early decades of the century and had encouraged its students to think beyond design issues to the broader challenges of corporate management and entrepreneurialism.[14] Compton strengthened the Division of Industrial Cooperation and Research when he assumed the presidency. He encouraged institute faculty to consult with industry, helping them to keep current on industrial technology and to supplement their incomes while also allowing them to place promising students with innovative employers.

Electrical engineering benefited, too, from the creative work being conducted elsewhere at the institute. Through the 1930s, researchers at MIT developed the Van de Graaff electrostatic generator, the differential analyzer, and the stroboscopic light. They discovered a method to detect radium in the human body, to send electrical transmission through human nerves, and to disperse complex waves of color spectra. They perfected new processes to liquefy hydrogen, to transmit radio waves through a hollow pipe (a precursor to the modern coaxial cable), and to apply ultrathin films to glass. Indeed, in 1940, when Compton issued a request to his department chairs for suggestions of possible areas where MIT could contribute to the war effort, John Slater from the physics department responded,

> There are a number of fields of practical importance in which our department is outstanding in the country. I may mention, for example, . . . automatic measuring and recording instruments . . . applied optics, electronics, X-ray structure . . . ultra-high frequency electromagnetic wave propagation, vibration problems, acoustics, thermodynamics, etc.[15]

Indeed, the great challenge for MIT in 1940 lay not in determining how it could contribute to the war effort but in how it could retain its teaching staff which was in danger of being poached by the Departments of War and Navy. Early in the war, James Killian, Compton's executive assistant (and later the institute's president), wrote somewhat desperately to the Office of the Undersecretary of War:

> The entire question of depletion of our faculty at the present time is one of grave concern, not only to the normal operation of the institute,

but also to our ability to fulfill the obligations which we have assumed toward the Army, Navy, government agencies, and defense industries. At the present time, 67 separate contracts for research and development are in force, 40 with government agencies and 27 with industrial concerns. Of the former the majority are of the highest priority and are designated as secret. The normal operative budget of MIT is approximately $3 million, while the increased load of defense researches approaches an additional $3 million superimposed upon it.[16]

Although MIT contributed to the war effort in many ways, its efforts in microwave radar were most significant. The ability to see through fog and darkness; to map incoming planes, ships, and torpedoes; to locate and guide distant objects; and to navigate without the aid of stars had profound ramifications for nearly all aspects of war. The potential importance in defense "can scarcely be overstated," wrote Harold Hazen, a professor of electrical engineering, to Compton.[17]

But the history of the Rad Lab is not so much a story of pathbreaking academic research as one of collaboration between qualified investigators and government administrators and funders. In June 1940, President Roosevelt convened a National Defense Research Committee (NDRC) to mobilize American science and technology for the coming war. Its top priority was to investigate microwave radar. With Alfred Loomis chairing a Microwave Committee, the NDRC spent the summer debating where to locate a proposed microwave laboratory and at a more basic level whether to hitch it to an industrial lab or a university. The committee initially considered locating the lab at the Department of Terrestrial Magnetism at the Carnegie Institution of Washington but found the physical space wanting, as well as being inconvenient to airfields and the ocean – both necessary for testing equipment on ships and planes. Moreover, several committee members felt that the lab should be situated in such a place as to have access to university scientists immersed in microwave research, as well as industrial engineers who had access to corporate financial resources, limiting the search only to Stanford and MIT. With urging from Vannevar Bush, the chairman of the NDRC and former dean of engineering at MIT, the committee explored the Boston area, where it could make use of a National Guard hangar in East Boston and leverage the prowess of MIT's electrical engineering department.[18]

The initial contract between the federal government and MIT was for $455,000 in October 1940. The deliberately misnamed Radiation Laboratory (it was always committed to research in microwave radar) was *in* MIT but not quite *of* MIT. That is, although the university administered the lab, the federal government selected its leadership and dictated its research program. Thus was established an important template for government–university cooperation in the sciences, in which federal funds would flow to university-based scientists, who maintained some of the freedom of academic researchers while being comfortably shielded from the pressures of consulting with industry or seeking outside foundation grants.[19]

The Rad Lab initially recruited a research team of 25 from among the electrical engineers who had been working at the Loomis Laboratory over the summer as well as from other universities. Within a year, the group had produced a workable microwave pulse radar system. By February 1941, the group had produced a set capable of being carried aboard an airplane. The staff grew rapidly, to 75 researchers by the end of 1940, to 500 by the end of 1941, and to 1,700 by the end of 1942. In 1942, also, MIT had constructed a purpose-built facility to house the lab's expanding team. The building was already too small by the time it was completed; MIT immediately added a five-floor addition and erected a temporary building to house overflow.

Functionality also increased exponentially. The first radar set, operating by February 1941, could track a small plane from 2.5 miles. By March, the range had increased to 5 miles, and by the end of the year to 20. The length of the microwaves dropped from 20 centimeters, to 10, to an extraordinary 3 by the war's end, allowing for blind bombing techniques so accurate as to preclude the axis powers from maintaining their wartime economies.[20] At the same time, the Rad Lab created its own manufacturing arm to produce the sets, with the first 50 delivered to the Army and Navy in 1942. The research program expanded rapidly during this time, branching into navigation, precision-bombing guidance, ground-controlled approach, and airborne fire-control equipment. By June 1943, the lab had delivered over 6,000 radar sets to the Army and Navy, with 22,000 on order.[21]

By the end of the war, the Rad Lab staff had grown to 4,000. Compton proudly boasted that it was the "largest research organization in the history of the world."[22] The lab, which had initially occupied 4,000 square feet of floor space, now occupied over 660,000 square feet. Its five-years contracts with the federal government had totaled just under $80 million, at a time when MIT's entire endowment was just about half that much. Radar sets designed and produced by the lab tracked enemy planes, ships, and torpedoes; guided aviation fire and bombing; detected incoming missile fire; guided planes to landing; and identified enemy movements.[23]

From a broader, institutional perspective, the Rad Lab established a template of cooperation between the federal government and independent universities which would define much of the nation's postwar scientific effort. The meshing of federal funding and direction, with academic independence and creativity created a powerful model for conducting scientific research. The federal funds allowed universities to conduct research at a scale that would otherwise have been impossible. During World War II, for example, MIT's total research staff doubled, its operating budget quadrupled, and its research budget (of which 85% came from the federal government) grew nearly tenfold.[24] At the same time, MIT maintained its independence and protected its permanent faculty, refusing to release professors to the military even under substantial wartime pressure. President Compton repeatedly thwarted efforts by the Army to poach the institute's faculty while defending the efficacy of university-based faculty conducting war-related research. The president wrote to General Charles Hines in 1942 that the MIT had become "a sort of 'happy hunting ground' for scientific and engineering talent during the war effort, and we can only hope to continue our own significant job in this war effort by retaining a few key men."[25] To department chairs, he wrote in a general memo,

> It is conceivable, for example, that a member of the staff might be called to duty with some outside agency on an important mission but that by accepting it he might render impossible the performance of some even more important activity by the institute.[26]

Fearful of losing Professor Charles Draper to the Army's officer corps, for example, Compton pointed out that Draper's work on the Sperry gun sight had been so valuable that the Navy had already ordered 21,000 of the new sights. At the time, Draper was directing an MIT course, which trained naval officers in fire control: a course not available at any other school in the country.[27]

In contrast to the centralized nationalization of scientific work underway in Germany at the time, MIT (and other universities) insisted that it was the very decentralization and independence of military research which made the US war effort so formidable. While German universities deteriorated in the years leading up to, and during, the war, US universities grew stronger. Frederick Keyes, a professor in MIT's chemistry department, presciently predicted in 1940,

> Assuming the United States and the Allies exploit their scientific and technical advantages fully, the Germans could be rapidly outclassed in improved implements of war. . . . To allow the country's scientific assets to fall under the control of government scientific bureaus means a close approximation to political domination with all that is implied in the stifling effect on ideas and the clogging of research progress.[28]

Far better, then, to allow the university-based science to continue unhindered while asking universities to contribute to the war effort through directed research and officer training. Through the war, MIT continued with its long history of training undergraduates for leadership positions through its officer training program and by training select groups of officers and industrialists in new techniques, technologies, and operations through special cohorts that rotated through the institute. In 1942 alone, for example, MIT trained 40 naval officers in naval construction and engineering and 20 in meteorology and mechanical and aeronautical engineering. At the same time, it took in 50 RCA employees to train them in radar usage.[29]

Vannevar Bush and the OSRD

The Rad Lab could hardly have achieved its aims absent a federal infrastructure to support scientific research. That infrastructure had its roots with the Scientific Advisory Board, established by President Roosevelt in 1933. Roosevelt envisioned the board in an advisory role to the White House and to that end invited Karl Compton to chair it, hoping that Compton could exploit his considerable scientific network to clarify scientific priorities for the executive branch. The board lacked a clear operational mission, however, and was disbanded two years later.[30] Through the 1930s, the White House lacked any scientific coordinating apparatus, leaving scientific research to the individual agencies (Agriculture, Commerce, War, Navy) to conduct in government labs and research stations. With the onset of World War II, however, the importance of coordinating military-related research increased. The Committee on National Defense, a loose amalgamation of the Secretaries of War, Navy, Interior, Agriculture, Commerce, and Labor, proposed a new NDRC to "correlate and support scientific research on the mechanisms and devices of warfare."[31] Notably, the new committee lacked operational authority to actually conduct research but rather merely extended the advisory and coordinating capacities of the Science Advisory Board. Roosevelt recruited Vannevar Bush, the former dean of engineering at MIT and then president of the Carnegie Institution of Washington, to chair the committee.[32] Other notables on the initial board included James Conant (president of Harvard), Compton, Frank Jewett (president of Bell Labs and of the National Academy of Science), Conway Coe (Commissioner of Patents); and Richard Tolman (professor of physical chemistry at the California Institute of Technology).

Within a year, it became clear that the advisory function of the NDRC was inadequate to the scientific challenges confronting a nation on the verge of war. In response, the President established a new Office of Scientific Research and Development (OSRD) under the Office of Emergency Management, answerable directly to the President. Moving Bush over to the directorship of the OSRD, the still-existing NDRC was placed beneath the OSRD as an advisory panel with Conant as chair. This two-step evolution of centralized control of the nation's research enterprise reflected the deep ambivalence of the White House to intervene in the scientific research for all but the most applied of military projects. In appointing Bush to chair the NDRC initially, Roosevelt explicitly directed him to coordinate his work with the Army and Navy but not to displace their operational functions. "It is not intended that the work of your Committee should replace any of the excellent work which these services are now carrying on, either in their own laboratories or by contract with industry," he wrote to Bush. "Rather it is to be hoped that you will supplement this activity by extending the research base and enlisting the aid of scientists who can effectively contribute to the more rapid improvement of important devices." The major asset that the committee had at its disposal was two officers detailed from the Army and Navy as liaisons – hardly the bulwark of power.[33]

By contrast, the OSRD was to have a budget (left undetermined at its outset), along with the power to contract directly with academic and industrial laboratories. In his executive order of June 28, 1941, Roosevelt explicitly articulated the mission of the OSRD to include mobilization of the scientific personnel and resources, supplementing scientific activities relating to national defense (including weapons development), and initiating and supporting war-related research in friendly countries.[34] Notably, the enabling order empowered the director of the OSRD to enter into contracts and agreements with individuals, universities, and industry.[35]

This was a new animal entirely, with a startlingly new portfolio. The director, sitting just beneath the President, commanded dotted-line reports from the Assistant Secretaries of Navy and War for Research and Development, as well as the director of the National Academy of Sciences, the NDRC, the National Advisory Committee on Aeronautics, and the Committee on Military Medicine, with a specially designated relationship with the British emissary for the coordination of research and development.[36] Endowed with the ability to forge contracts, and with high priority within the President's war cabinet, the OSRD would grow in size and importance over the following four years.

Vannevar Bush, the President's point man on scientific affairs, was born in Everett, Massachusetts in 1890, son of a Universalist minister. Always a tinkerer, he registered a patent for a surveying machine while still an undergraduate. After college at Tufts, he earned a doctorate in electrical engineering from MIT in 1916, and spent the following three years as an engineer at the American Radio and Research Corporation (ARR) while also working as a consultant with the American Research and Development Corporation (AMRAD) creating advanced radio devices. It was through his work at ARR and AMRAD that Bush gained appreciation for the synergistic roles of academic and industrial research. During World War I, AMRAD detailed Bush to the US submarine headquarters in New London, Connecticut, where he worked on anti-submarine research. In 1919, he returned to MIT as an associate professor of electrical power transmission and began a rapid ascent through professorial and administrative ranks. Placed in charge initially of the introductory course in electrical engineering at the university (he published a textbook in the field in 1922), he was quickly made director of graduate studies and research within the department. By 1932, he was dean of engineering at MIT, which effectively made him the institute's second most powerful administrator.

Initially a researcher of radio technology, Bush expressed an early interest in computers, devising an analog calculating device to abet him with his own calculations. Through the 1930s, he worked to perfect a differential analyzer, which could be used to solve complex systems of equations. By 1935, he had produced a punch-tape driven device – the Rockefeller

Differential Analyzer – which included 200 electronic tubes and weighed 100 tons. It was used extensively by the Navy during World War II to calculate firing trajectories and placement of radar antennas.

In 1938, Bush became president of the Carnegie Institution of Washington, an independent set of laboratories endowed by Andrew Carnegie which occupied offices and laboratory buildings around Washington DC. He would maintain his position with Carnegie over the following 16 years, even while holding positions as the chair of NDRC, the director of OSRD, and a board member of American Telephone and Telegraph. Although he maintained a lifelong interest in technical innovation, his true gifts were administrative – the ability to negotiate agreement among strong-minded people to advance a set of complex goals and to communicate those goals persuasively to diverse constituencies. He wrote incessantly and fluidly, producing detailed missives of 10–15 pages in a few hours describing a set of executive goals, a political strategy for moving legislation or budgets forward, or a recruitment strategy for a particularly elusive researcher or university dean. Born to modest means, he worked comfortably among wealthy and powerful individuals, relying on his disciplined analytical faculties and use of persuasive language to bring others to his side. Of note, he refused the title of "scientist," describing himself as an engineer throughout his life.[37] He admitted in the decades after World War II that the line between the disciplines was blurring. "Engineers, those who are in the forefront of advance, are becoming more entitled to be recognized as scientists in their own right. Applied scientists, under the pressure of war and its aftermath, have often become accomplished engineers as well."[38]

In taking on the OSRD role, Bush committed himself as much to a guiding philosophy as to a specific set of research objectives. Concerned that the Army and Navy's historic control over their own munitions and systems research would constrain their ability to support weapons innovations, he fought successfully to wrest the monopoly on such research from military laboratories (performed by full-time military personnel) and instead to outsource it to independent academic engineers and physicists in university and industrial labs.[39] In support of this effort, he worked to perfect the government research contract, under which the government would demand a certain level of effort along broadly conceived lines of inquiry without exactly dictating a deliverable product. That is, scientific "work" itself had to be reconstrued as a designated effort which might or might not produce a workable solution to a military problem. As part of the development of these contracts, Bush committed the OSRD to reimbursing host universities and industrial firms for soft, indirect costs associated with a specific investigator's research. To be successful, OSRD contracts could not be viewed by university administrators as imposing a cost on host institutions.

In an effort to further buttress this new partnership, Bush sought to create more trust between civilian scientists and military leaders; relationships which would evade the hierarchical chains of command with which military officers were comfortable while pushing civilian researchers to direct their efforts to problems of interest to the military. This dance needed to be carefully choreographed, lest scientist or officer abandon the effort in frustration. Ultimately, these cooperative relationships led to several of the most important innovations of the war, including the radar and the atomic bomb.[40] As part of this effort, Bush sought to put in place security measures tight enough to satisfy security-minded military leaders but not so cumbersome as to antagonize more liberal-minded scientists.

Closely associated with these goals was Bush's efforts to loosen some of the traditional restrictions on military advancement. Experience in combat leadership had historically been the path to promotion for career Army officers. Bush believed strongly that innovative weapons technology was as important as bravery, tactics, and troop strength in winning a war and that Army and Navy promotion standards needed to allow for key placement of technically minded officers who might lack the combat experience which underlay the traditional path to promotion. If cooperation between soldier and scientist was the key to victory, then Bush needed for the Army to retain officers who tolerated, if not actually appreciated, scientists. Officers needed to accept scientists as "equals and independent in authority, prestige, and in funds."[41]

Bush and Conant were scientific and academic elitists, having served for many years in positions of academic leadership in the nation's most prestigious universities, where they sought out the best scientific and engineering talent available in service of teaching and research. Both brought their outlook to their work for the government. As chair of the NDRC, Conant created a comprehensive catalogue of the nation's most accomplished physicists and chemists who could be available for wartime service through contract. This catalogue became the template through which the OSRD directed funds to universities and industrial concerns over the next five years, ensuring that the most innovative and accomplished scientists were called to war service but having a secondary effect of consolidating academic strength in the universities which were already at the forefront of research before the war. That is, under the accelerant of OSRD and military funding, the strongest universities became stronger still during the war, using federal funds to recruit new talent, build new laboratories, and expand already healthy academic programs. Exacerbating the pattern was the fact that OSRD and military contracts favored universities which possessed adequate facilities to carry on large research projects while also periodically asking universities to "front" funds for lab development until budget measures could be approved by Congress. Both of these policies tended to concentrate federal funds in universities which had already been academically prominent before the war. Moreover, both the OSRD and the Army preferred consolidation – one large lab as opposed to multiple coordinated smaller ones. This was not so much a scientific bias as an administrative one, as it was easier to channel research funds through one powerful lab director rather than coordinating the work of several less powerful ones.

The OSRD tended to create regional labs, usually based at a particular university (albeit sometimes the creature of a regional syndicate of universities), with one powerful and wealthy central university to front the money, build the labs, and recruit the talent. President Compton, of MIT, recalled after the war:

> To meet this emergency, our Executive Committee, each year, voted to underwrite the government, so to speak, to the extent of a half million dollars in order that we could, where necessary, assure our research staff of appointment in the coming fiscal year. An anonymous distinguished philanthropist, learning of our embarrassment in this matter, generously agreed to supplement our underwriting by an additional half million dollars. We were thus enable to hold our staff, and fortunately, these underwritings were not called upon because Congress ultimately voted the desired appropriations and our contracts were renewed.[42]

The OSRD budget would grow exponentially during the war, along with research budgets of the Army and Navy. In 1944, the combined research budgets of the three agencies was $628 million, with another $860 million spent on the effort to develop an atomic bomb, for a combined total military research budget of nearly $1.5 billion for that one year. In 1939, by contrast, the entire combined research budgets of the nation's armed services had been $13 million. "Never once did we ask for funds and fail to secure them promptly," claimed Bush.[43] By the end of the war, this torrent of funding had fundamentally altered the relationship of the federal government with the nation's scientific enterprise, and had, in fact, altered the scientific enterprise itself. Whether the funding should continue in the absence of a military adversary became a compelling question for postwar planners.

The Manhattan Engineering District

Along with radar, the atomic bomb, developed over a frenzied four-year period of research and construction, marked the beginning of government-funded "big" science. The project required constructing new cities, large laboratories, and the massive equipment needed to produce plutonium and to separate the fissionable U^{235} isotope from the inert U^{238}, which dominated naturally occurring uranium deposits. By the end of the war, the Army and the OSRD had spent $2 billion on the project (not all for scientific research) – a staggering sum on weapons development which dwarfed all US scientific ventures until the Apollo space program in the 1960s.[44]

Nuclear energy, the energy locked within the nucleus of the atom, was discovered by Henri Becquerel in France in 1896, when he observed that uranium ore emitted energy which could register on a light-sensitive photo emulsion. This naturally occurring radiation hinted at hitherto unknown forces, but the energy appeared to be immune to capture. The key to manipulating it proved to be the neutron – the proton's neutrally charged counterpart in the nucleus, discovered by James Chadwick in 1932 – which could enter a nucleus unrepelled by the proton's positive repellant force and thus destabilize the "strong forces" which held the nucleus together. The destabilized nucleus would sometimes split apart entirely, converting part of its mass into pure energy while releasing more neutrons to enter neighboring atoms and thus create a self-sustaining fission reaction. Isidor Rabi, a prominent mid-20th-century physicist, compared the effect of a neutron entering a nucleus to be "as catastrophic as if the moon struck Earth."[45]

Neutrons were cheap and available. By the late 1930s, physicists could purchase radium salt from commercial vendors and by mixing it with beryllium produce a reliable stream of neutrons. E. O. Lawrence's new cyclotron accelerator, which he developed in his lab at Berkeley in 1935, provided an important venue for neutron researchers – one in which neutrons could be speeded up and given some guidance. In 1938, Otto Hahn and Fritz Strassman, working in Germany, discovered that certain uranium atoms (specifically the rare U^{235} isotope) produced two barium atoms upon bombardment with a neutron, a discovery which Lisë Meitner and Otto Frisch, working at the Kaiser Wilhelm Institute, concluded indicated the splitting of the uranium atom. At nearly the same time the Italian physicist Enrico Fermi, working at labs first at Columbia University and later at the University of Chicago, experimented with inducing fission by using a radium-beryllium source to fire neutrons at uranium, producing greater numbers of neutrons – an indication of the possibility of a sustained fission process. Over the next two years Fermi built an atomic "pile" of graphite blocks bored with holes to allow the introduction of tubes of enriched uranium pellets. By manipulating the size of the pile, the placement of the uranium columns, and the depth of a cadmium-coated control stick, Fermi produced the world's first sustained fission reaction in December 1942, in a squash court underneath the University of Chicago's football stadium.

By applying Einstein's special theory of relativity (published by the German physicist in 1905), physicists could deduce that the amount of energy released by splitting a uranium atom was potentially gargantuan. To put it in perspective, a typical exothermic chemical reaction between two reagents produced 5 electron volts of energy, while the comparable energy release in a nuclear reaction with the same mass of reactants was 200 million electron volts – 40 million times as much. In theory, a stock of enriched uranium the size of a baseball could release as much energy as a freight train worth of coal. Rabi wrote that when Fermi learned of the Hahn–Strassman experiments, by nightfall he was "already speculating on the size of the crater which would be produced if one kilogram of uranium were to disintegrate by fission."[46]

In the Fall of 1939, the Hungarian-born physicist Leo Szilard drafted a letter to send to President Roosevelt under Einstein's signature in an effort to make the President aware of the potential of a new fission-powered weapon and to warn him of progress being made by German scientists at the Kaiser Wilhelm Institute. Roosevelt responded by establishing an Advisory Committee on Uranium, chaired by Lyman Briggs, the director of the National Bureau of Standards, to investigate the issue and advise. The following year the President moved the Uranium Committee to the newly formed NDRC where Vannevar Bush could integrate its work into other war-related scientific work. (Roosevelt had explicitly mentioned the recent discoveries in the field of "atomistics" and their possible relation to defense concerns when he had invited Bush to chair the NDRC.)[47] In May 1941, a committee of physicists at the National Academy of Sciences recommended to the NDRC that the government make a "strongly intensified effort" on uranium research.[48] Later that year, the OSRD recommended to Roosevelt that the Army begin an all-out effort to develop a fission bomb. In June 1942, the Army created a special engineering unit, originally called the DSM project (Development of Substitute Materials), which officially became the Manhattan Engineering District (MED) that August. On September 17, Major General Leslie Groves of the Army Corps of Engineers was placed in charge of the project.[49]

The challenge facing Groves was not how to build a bomb, but rather how to produce an adequate amount of fissionable material to produce a sustainable reaction. The fissionable isotope of uranium, U^{235}, constituted just over one-half of 1 percent of naturally occurring uranium ore. A workable bomb would need a core of at least 80 percent U^{235}. Separating the U^{235} from the slightly heavier U^{238} was tremendously difficult, but in theory it could be done using one of four techniques. In each approach, the uranium ore would be gasified so that the individual atoms could move easily within a container or matrix. Then, the U^{235} isotopes could be separated out by using either one of the following: a gaseous diffusion process, a centrifuge process, a thermal diffusion process, or an electromagnetic process. Each of the processes exploited the slight difference in mass between the two isotopes. None of the approaches could be easily achieved, and all would involve substantial investment in precisely engineered equipment. Centrifuges, for example, would need to spin at tens of thousands of revolutions per minute to create a high-enough centripetal force to selectively force the U^{238} atoms to the perimeter of the device. At such speeds, bearings and axles would need to be precisely tooled, lest the machines tear themselves apart.

Another approach was not to separate out U^{235} at all but rather to aggregate a by-product of fission, a new man-made element called plutonium (after the ninth planet), which physicist Glenn Seaborg had first produced using Lawrence's cyclotron in the Berkeley lab in 1940. Seaborg had discovered that when uranium was split, at least some of the excess neutrons were absorbed by the inert U^{238}. The now unstable nucleus emitted two electrons, converting two neutrons into protons and creating a new element slightly heavier than uranium and equally fissionable. Plutonium was a by-product of Fermi's fission pile and could potentially be produced in bulk, leading to a cheaper, faster method of producing adequate fissionable material. Grove and his senior planners, unsure of which approach was most likely to yield a workable result, chose to pursue all simultaneously.[50]

Groves's task was immense – to produce an adequate supply of highly enriched fissionable material to build a bomb and to figure out a way to construct and transport such a bomb in a manner as to forestall the beginning of the fission reaction until detonation. This last part was trickier than it might seem, for so rapidly could a chain reaction transpire that a mass of fissionable material could blow itself apart before it had time to achieve comprehensive fission. It was almost as if the fuse burning down to a dynamite stick ignited just a small portion of the dynamite, which promptly blew apart the rest of the explosive before it had time to detonate. While the problem was eventually solved in two different ways (and thus the different designs of the two bombs dropped on Japan), it was never solved entirely. Over 98 percent of the fissionable material in the bombs dropped on Japan failed to undergo fission.

Groves situated the work of the MED largely in three locations: in Oak Ridge, Tennessee, where the Clinton Engineer Works built gaseous and thermal diffusion and electromagnetic separation plants to separate out uranium isotopes; in Richland, Washington, where the Hanford Engineer Works on the Columbia River produced plutonium; and at Los Alamos, New Mexico, where the bomb itself was developed and assembled.[51] Each of the three installations would ultimately constitute a small city. At Oak Ridge, for example, workers used 200 million board feet of lumber, 400,000 cubic yards of concrete, and 50,000 tons of iron and steel to build 10,000 houses and dormitories, 16,000 barracks placements, and a variety of supermarkets, hospitals, restaurants, and even movie theaters to house a population, which by the war's end reached 78,000, in addition to constructing the factories for the thermal, magnetic, and gaseous diffusion processes. At Hanford, engineers moved 25 million cubic yards of earth to build the world's largest earthen dam (to provide coolant water for the plutonium-producing reactor), laid 400 miles of electrical transmission lines, paved 345 miles of roads, and poured 780,000 cubic yards of concrete to build and furnish the dwellings and workspaces for the 45,000 people who would relocate there over three years.[52] Los Alamos, the brain center of the project, never grew as large, but it ultimately housed over 4,000 scientists and support staff. Whether measured by total budget or man-hours, the scale of the MED was the equivalent of building a Panama Canal each year for three years.[53]

This extraordinary venture produced the first successful nuclear explosion at Alamogordo, New Mexico (Trinity test site) on July 16, 1945.[54] Three weeks later, the United States dropped a 15-kiloton uranium bomb on Hiroshima, Japan, followed three days later by a 20-kiloton plutonium bomb on Nagasaki. Both bombs devastated the core part of their target cities and killed over 100,000 in the initial shock waves and heat blasts. (Many thousands more would die in the weeks after from radiation sickness and burns.) A journalist invited to ride in one of the two accompanying planes on the Nagasaki bombing run wrote:

> Observers in the tail of our ship saw a giant ball of fire rise as though from the bowels of the earth, belching forth enormous white smoke rings. Next they saw a giant pillar of purple fire, 10,000 feet high, shooting skyward with enormous speed.
>
> By the time our ship had made another turn in the direction of the atomic explosion the pillar of purple fire had reached the level of our altitude. Only about 45 seconds had passed. Awe-struck, we watched it shoot upward like a meteor coming from the earth instead of from outer space, becoming ever more alive as it climbed skyward through the white clouds. It was no longer smoke, or dust, or even a cloud of fire. It was a living thing, a new species of being, born right before our incredulous eyes.[55]

Six days later, the Imperial Government of Japan surrendered.

The Manhattan Project represented a new level of cooperation between the government and scientists. No laboratory on Earth could have come close to the total output of work and discovery that the MED had accomplished in just three years without the backing of the world's most powerful government making an unprecedented financial commitment to the project. At the height of the MED, Groves's power was absolute; it was as if he commanded a small state. He described his powers in retrospect: "No officer I ever dreamed of had the free hand I had in this project; no theater commander ever had it and I know of no one in history who has had such a free hand."[56]

But Groves could have accomplished little without the cooperation of independent universities. The Los Alamos laboratory, where the entire project was planned and overseen, was managed as an arm of the University of California, directed by the brilliant young physics professor J. Robert Oppenheimer. Although initially met with some skepticism by senior military officers, Oppenheimer directed the work of thousands of physicists, chemists, metallurgists, and nuclear engineers to succeed in building two workable devices based on different principles of material extraction in just three years. Although abrasive and eccentric, he successfully shielded his research and development team from the sort of rigid bureaucratic thinking that had undermined previous arms and ordnance projects. Although demonized later for his Communist affiliations and ultimately stripped of his security clearance during the Manhattan Project, he proved to be an adept leader – authoritative, inspiring, and flexible as needed. He described the *esprit de corps* of Los Alamos after the war:

> I have never known a group more understanding and more devoted to a common purpose, more willing to lay aside personal convenience and prestige, more understanding in the role that they were playing in

their country's history. Time and time again we had in the technical work almost paralyzing crisis. Time and again the laboratory drew itself together and raced new problems and got on with the job. We worked by night and by day; and in the end the many jobs were done.[57]

The experiences of wartime science produced new models of doing and funding science. Under pressure to produce innovative tools and munitions rapidly, the federal government turned to the nascent but largely untapped creativity, expertise, and inquisitiveness in the nation's universities and industrial labs and asked that they work with the government toward national priorities. Through the coordinating agency of the OSRD, and working closely with directors of research for the uniformed services, universities across the nation diverted their faculty toward research goals dictated by military necessity while at the same time shielding those faculty from the narrow and hierarchical thinking intrinsic to large government (and military) bureaucracies. The process had been extraordinarily expensive and had left federal budget masters to navigate the uncertain waters of contractual accountability, but in the end the relationship proved highly successful. Scientists at the Rad Lab liked to boast later, "Radar won the war; the atom bomb ended it."[58] The question that lay for a president turning his administration to a peacetime footing was whether or not the relationship could be, or should be, sustained in the absence of a clear and present threat.

Notes

1 "GCA," MIT, 154: NDRC, p. 1.
2 Ibid., p. 4.
3 See Daniel Kevles, "Cold War and Hot Physics: Science, Security, and the American State, 1945–56," *Historical Studies in the Physical and Biological Sciences*, 20:2, 1990. Also, Lee DuBridge, "Policy and the Scientists," *Foreign Affairs*, April 1963, for a discussion on the overall contribution of physicists and engineers to the war effort.
4 "Bombing Through Overcast," MIT, 154: NDRC.
5 Steven Dick, "John Quincy Adams, the Smithsonian Bequest, and the Founding of the U.S. Naval Observatory," *Journal for the History of Astronomy*, 22:1, 1991. Adams would live long enough to celebrate the creation of the United States Naval Observatory in 1844, writing,

> There is no richer field of science opened to the exploration of man in search of knowledge than astronomical observation; nor is there, . . . any duty more impressively incumbent upon all human governments than that of furnishing means and facilities and rewards to those who devote the labors of their lives to the indefatigable industry, and unceasing vigilance, and the bright intelligence indispensable to success in these pursuits.
>
> From John Ventre, "JQA's Role in American Astronomy," http://buhlplanetarium2.tripod.com/bio/jqa/astrorole.html, 7/22/20

6 Quoted in Lee Dubridge, "Science and Government," *Chemical Engineering News*, 31:14, 1953, p. 1386.
7 Daniel Kevles, "Scientists, Arms, and the State: J. Robert Oppenheimer and the Twentieth Century," in Cathryn Carson and David Hollinger, eds., *Reappraising*

Oppenheimer: Centennial Studies and Reflections, University of California Press, 2005, p. 330.

8 See Michael Sherry, *Preparing for the Next War: American Plans for Postwar Defense, 1941–45*, Yale University Press, 1977, pp. 122–23.

9 "Background of Research in the United States," NACA/IDC, 1: ICSRD Reports.

10 The schools with and the number of publications were MIT (15); Johns Hopkins (11); Stanford (9); Cornell (5); and Purdue (4). Statistics from Frederick Terman, "The Electrical Engineering Research Situation in the American Universities," *Science*, 65:1686, 1927.

11 Ibid., p. 2.

12 Ibid.

13 For a lengthy description of Loomis's work, see Jennet Conant, *Tuxedo Park: A Wall Street Tycoon and the Secret Palace of Science That Changed the Course of World War II*, Simon and Schuster, 2003.

14 See S. S. Schweber, "Big Science in Context: Cornell and MIT," in Peter Gallison and Bruce Healy, eds., *Big Science: The Growth of Large-Scale Research*, Stanford University Press, 1992. Schweber writes, "To a large extent, MIT's impressive growth during the decades around the turn of the century resulted from its success in developing a curriculum that made its graduates not only successful engineers but also able managers." p. 151.

15 Slater to Compton, June 5, 1940, MIT, 152: National Defense.

16 Killian to Kohloss, June 20, 1941, MIT, 152: National Defense.

17 Hazen to Compton, June 10, 1940, MIT, 152: National Defense.

18 Guerler to Killian, August 17, 1945, MIT, 154: NDRC Rad Lab.

19 "Selection of the MIT as the Site for the Radiation Laboratory," MIT, 154: NDRC Rad Lab. Bush describes a disagreement between Karl Compton and Frank Jewett, the president of the Bell Telephone Laboratory, and a member of the NDRC in locating the Rad Lab. He remembered,

> Under Compton, there was assembled at the Radiation Laboratory probably the hottest crowd of physicists on electronic gadgetry and the like that was ever put together. A lot of them were prima donnas; there were a couple of thousand people in the laboratory and it was quite a show. Frank Jewett had been the fellow who had really put the Bell Laboratories together, and he'd directed it. It was the pride of his whole career. It was just too much for him to think that a crowd of youngsters could do something better than the Bell Labs could.
> Quoted in James Killian, *The Education of a College President*, MIT Press, 1985, p. 24

20 See Killian, *The Education of a College President*, p. 26.

21 "Radiation Laboratory," MIT, 154: NDRC Rad Lab.

22 "Radar and the Atomic Bomb – Two Largest Scientific War Projects," MIT, 154: NDRC Rad Lab.

23 "Radiation Laboratory Statistics," MIT, 154: NDRC Rad Lab.

24 Paul Forman, "Behind Quantum Electronics: National Security as Basis for Physical Research in the United States, 1940–1960," *Historical Studies in the Physical and Biological Sciences*, 18:1, 1987, pp. 156–57.

25 Compton to Hines, March 24, 1942, MIT, 152: National Defense.

26 Compton, "To the Members of the Staff," MIT, 152: National Defense.

27 Compton to Bundy, February 24, 1942, MIT, 152: National Defense.

28 Keyes to Compton, June 6, 1940, MIT, 152: National Defense.

29 Compton, *Report of the President*, 1942.

30 The SAB extended the work of the National Academy of Sciences and the National Research Council, established during the Civil War and World War I with similar

goals of advising the President and Congress on war-related scientific developments. Both still exist.

31 From J. A. Furer, "US Naval Administration in the Second World War," AW, 321: OSRD.

32 Andrew Carnegie endowed at least 20 philanthropic and civic organizations with his fortune, most of which bear his name in the title. The Carnegie Institution was specifically devoted to scientific research in five different fields, including geophysics, astronomy, plant biology, and terrestrial magnetism, all conducted in labs located in and around Washington DC. The institute was renamed the Carnegie Institution for Science in 2007, in an effort to distinguish it from the many other foundations and institutions bearing the Carnegie name.

33 FDR to Bush, June 15, 1940, MIT, 153: NDRC.

34 Executive Order 8807 Establishing the OSRD in the Executive Office of the President, June 28, 1941.

35 Ibid.

36 Organization charts can be revealing. The initial chart of the OSRD showed it lying above all of its dotted-line reports and just below the Office of Emergency Management. The relationship with the British scientific emissary was explicit. The original can be found in FDR, PSF, 143: NDRC.

37 See Jerome Wiesner, *Vannevar Bush*, National Academy of Sciences, 1979, p. 102.

38 Quoted in Schweber, "Big Science," p. 175.

39 This crucial need was perceived by Bush even before taking over the OSRD. In 1940, he wrote to Frank Jewett about an upcoming visit to Wright Field in an effort to persuade the Army to transfer some of its research efforts to universities. Bush to Jewett, June 15, 1940, MIT, 153: NDRC.

40 Lee DuBridge, the director of the Rad Lab, makes this argument in "Policy and the Scientists," *Foreign Affairs*, April 1963, p. 574.

41 Quoted in Michael Sherry, *Preparing for the Next War*, Yale University Press, 1977, p. 136.

42 "Selection of OSRD Contractors," MIT, 154: NDRC Rad Lab.

43 Quoted in Gregg Zachary, *Endless Frontier*, The Free Press, 1997, p. 289.

44 The bibliography is huge, but an excellent and readable overview of the history is Richard Rhodes, *The Making of the Atomic Bomb*, Simon and Schuster, 1986.

45 I. I. Rabi, "The Physicist Returns from the War," *The Atlantic*, 1945, pp. 111–12.

46 Ibid., p. 112. Not surprisingly, there is some disagreement over the precise sequence of events which led to the development of a controlled fission reaction. In 1954, three French scientists – Joliot, Alban, and Kowalski – sued the US AEC for patent violation based on their contributions to the development of atomic energy prior to 1941. Joliot had discovered artificial radioactivity in 1934, and the three men, together, published the first report of neutron emission in *Nature* in 1939. Further papers buttressed the scientists' claim on much of the early work on controlled and sustained fission. Ultimately, the AEC, in conjunction with the US State Department, reached a monetary settlement with the French scientists. In 1966, the AEC's Fermi Award was given to Hahn, Strassman, and Lisë Meitner for their early work on fission. It is safe to say that individuals situated in multiple labs around the world in the 1930s were working on problems related to controlling, sustaining, and measuring fission reactions. By 1939, to use one data point, nearly 100 articles on nuclear fission had already been published in peer-reviewed scientific journals from around the world. Thus, the exact "inventor" or "discoverer" of fission is hard to assign. For details on the case involving the French scientists, see "Proposed Settlement of Application for Award by Commissariat a L'Énergie Atomique and Messrs. Joliot, Halban, and Kowarski," LBJ, WHCF, FG202, 263:8/4/67–12/31/67.

47 FDR to Bush, June 15, 1940, MIT, 153: NDRC.

48 Quoted in Robert Norris, *Racing for the Bomb*, Skyhorse Publishing, 2002, chapter 10.

49 The decisions to build and drop the bomb and the precise credit due to different individuals are fraught topics which have been the subject of much scholarly debate. For a more skeptical take on these events, see Stanley Goldberg, "Climate of Opinion: Vannevar Bush and the Decision to Build the Bomb," *Isis*, 83:3, 1992.

50 See Glenn Seaborg, "The First Nuclear Reactor, The Production of Plutonium, and Its Chemical Extraction," *International Atomic Energy Bulletin*, December 2, 1962, p. 1.

51 The MED used equipment at numerous other locations in the United States, including a major electromagnetic separation plant at Berkeley, but these three were the most substantial.

52 Details from a War Department memo, "Background Information on Development of Atomic Energy Under Manhattan Project," HST, SRB, 5: AEC.

53 From "Construction for Atomic Bomb Production," *Engineering News Record*, December 13, 1945, p. 1.

54 Bush, who was present at the test site, reported thinking, "If this thing goes off with a hell of a lot more force than we've calculated, they'll have to get a new [OSRD] head." Quoted in Zachary, *Endless Frontier*, p. 279.

55 "Eye Witness Account Atomic Bomb Mission Over Nagasaki," HST, PSF, 174: Atomic Bomb: Cabinet: James Byrnes.

56 From Barton Bernstein, "Reconsidering the 'Atomic General': Leslie R. Groves," *Journal of Military History*, 67, July 2003, p. 900.

57 S.S. Schweber, *In the Shadow of the Bomb*, Princeton University Press, 2000, p. 104.

58 Richard di Dio, "The Scientist-Tycoon Whose Work Helped Win WWII," *Philadelphia Inquirer*, July 21, 2002.

3 Postwar Realignment and the Office of Naval Research

The Bush Report

In October 1944, Oscar Cox, an administrator of the Foreign Economic Administration, suggested to Vannevar Bush that President Roosevelt appoint a committee to prepare a planning document for American science after the war's conclusion. The committee would report on the effects that the government's scientific efforts had wrought on the conduct of the war and on the economy and would lay out a strategy to maintain strength in basic research in the coming decades.

Roosevelt responded with Executive Order 9791 to create just such a committee and named Bush chair and primary author. The order asked Bush to consider a number of tricky challenges, including balancing dissemination of basic research with the reasonable requirements of national security; continuing successful wartime scientific enterprises; providing government scientific aid to private universities within the framework of continued research in government labs; and identifying and developing future scientific talent.[1]

The report, "Science, The Endless Frontier," was delivered to President Truman in July 1945. In stern language, the report asserted that virtually all wartime technological breakthroughs (radar, proximity fuses, the atomic bomb) were built on primary (also called "basic" or "fundamental") research that had been conducted before the war, which had constituted a trove of scientific "capital." Of concern was that the bulk of this capital had been produced in Europe before the war and that the United States was in the process of depleting the pool of capital while failing to add new innovative discoveries to renew it.[2] "The intellectual banks of continental Europe, from which we have borrowed, have become bankrupt through the ravages of war," Bush wrote. "In the next generation, technological advance and basic scientific discovery will be inseparable; a nation which borrows its basic knowledge will be hopelessly handicapped in the race for innovation."[3]

To renew this pool, Bush proposed a broad framework of government support for basic research, predicated on substantial government support of *individuals* (rather than institutions) conducting research in private and nonfederal state-funded institutions. These institutions – mostly the nation's

DOI: 10.4324/9781003363897-3

private and state colleges and universities but also some private research institutes – would be able to leverage this government support to recruit and retain better faculty on which they could build stronger academic programs to train the next generation of scientists.

To dispense these funds, Bush recommended the creation of a new National Research Foundation, consisting of presidentially appointed members who were not otherwise connected with the federal government ("of broad interest in and understanding of the peculiarities of scientific research and education"). The members would be given great discretion and freedom to allocate funds to the most promising and innovative researchers around the nation with the counsel of the National Academy of Sciences. The new research foundation would allocate funds through five divisions: medical research, natural sciences, national defense, scientific personnel and education, and publications and collaboration. Substantial funding should also be made available to support students at the undergraduate and graduate levels, as well as for postdoctoral researchers and early stage scientists.

Of particular importance to Bush were two caveats: the preservation of the autonomy of the individual universities which would host the funded scientists, and long-term (five years or more) commitments to specific projects to give the researchers the time needed to take on large projects. Bush, with his years of experience as an MIT professor and dean, believed that institutional autonomy was a critical factor in fostering scientific excellence and innovation. He wrote, "Support of basic research . . . must leave the internal control of policy, personnel, and the method and scope of the research to the institutions themselves. This is of *utmost* importance." He also emphasized that fiscal uncertainty could, *would*, undermine even the most gifted and committed scientists' work. He admonished his audience, "Basic research is a long-term process – it ceases to be basic if immediate results are expected on short-term support."[4]

Congress responded by holding hearings on the report to consider next steps. The primary exhibit, Bush himself, presented to a joint session of the Committees on Commerce and Military Affairs in the Senate that October. Bush was savvy enough to understand that the most compelling argument for future government support of science lay in the great wartime achievements he had achieved through his work at the OSRD. He lectured the assembled senators, "Wise federal support for scientific research is just as essential to the successful prosecution of the war against disease and for our general progress and prosperity as it is vital to the successful conduct of war."[5] Advocating for his proposed research foundation, he emphasized the great paucity of scientific talent in the nation. The nation had always undertrained its promising young men and women and failed to identify and support talent.[6] "There are many reasons for this," he told the committee. "The most important is that much good talent has failed, because of the fortuitous circumstances of family fortune." During the war the nation had trained few engineers or scientists for professional research, having "consistently put in uniform young men who could have more effectively served the nation in our industries and laboratories."[7]

His central point, however, remained the need to center future scientific efforts on scientists working at independent public and private universities. Having seen the extraordinary success of regimented, hierarchical, government science ("an autocratic form of organization") during the war, he rejected it as a workable model for innovative inquiry. True, the huge laboratories of the Manhattan Project had been able to produce the weapons demanded of them, but this was merely a tactic in an all-consuming war, warned Bush, not a path to advance the frontiers of knowledge. The nature of bureaucracy, particularly one grounded in the military, was to constrain creativity. Only in universities could "free, inquiring minds find their greatest encouragement and inspiration." In the absence of a steady flow of government research funds, these institutions could ill-afford to retain scarce scientific talent, which would soon be siphoned off by wealthy industrial concerns. "It is not so much the salaries which industry offers," Bush told the senators, "as it is the facilities and funds for research which attract the scientist away from the colleges and universities."[8]

So the message was trifold: government support; scientific independence; and developing the next generation. Other witnesses concurred. Detlev Bronk, a professor at the University of Pennsylvania and later the president of the Rockefeller Institute and of Johns Hopkins University, urged senators to support research, "for it is from scientific research that our citizens have the greatest promise of higher standards of living, better health, and security against the dangers of foreign aggressions."[9] Like Bush, he exhorted senators to consider the effects of inconsistent and inadequate funding streams on scientific progress, undermining the "economic status of the scientist" and driving science into the province of "part-time avocation of an overworked teacher." "Everywhere," explained Bronk, "the scientist is notoriously underpaid." A doctorally trained researcher with several years of postdoctoral experience had a mean starting salary of $3,000.[10] Years of a universal draft had pulled promising students from the classrooms for wartime service, derailing the lengthy education necessary for the preparation of effective scientists. "Our scientific and technical manpower is seriously deleted as a result of our policies during the war," rued Bush. "We placed altogether too many men of special training in the armed camps."[11]

Response to the Bush report was largely positive, particularly from active scientists. While the research foundation would go through several iterations before emerging successfully from the legislative process five years later, the concept of such a foundation was barely opposed. A group of prominent scientists and research administrators met in New York, in January 1946, to form an *ad hoc* committee to support the report with the aim of lobbying for one of the proposed bills then circulating in Congress.[12] It was difficult to push back against more money, more autonomy, more scientific freedom, and greater empowerment of universities. Perhaps the only institution which consistently questioned the wisdom of the idea was the one that had, until that moment, funded the bulk of scientific research over the previous decade: the US military. The argument was not really about science, *per se*, but rather about control. Could a community of iconoclastic, free-thinking researchers, unbridled by hierarchy and largely unaccountable, produce the new knowledge the nation would need as it turned its attention elsewhere?

The Steelman Report

The end of World War II forced a realignment of national priorities. The swollen Army and Navy were quickly demobilized, even as the United States faced the threat of an increasingly aggressive Soviet Union. The nation invested in houses, highways, and schools. Industrial firms pivoted to a peacetime footing, hiring discharged servicemen to produce goods and services for a rapidly growing consumer base.

But what of science? Bush's report from 1945 provided a general template for basic research, but it failed to offer detailed direction on the disposition of the massive wartime scientific buildup. What should happen to Los Alamos? Hanford? Oak Ridge? How should universities, their research faculty swollen through wartime contracts, plan for the coming decade? And, most importantly, should research recede to prewar levels and norms or build off of the great successes of the war?

In October 1946, Truman created a Presidential Research Board to be directed by his close advisor John Steelman. Steelman had served in a variety of senior administrative rolls for the President – at the time he was officially the White House reconversion director – and in Truman's second term would take on the executive tasks which would later devolve to an official White House Chief-of-Staff.[13] A nonscientist (trained in political science and administration), his principal qualification for the position was the deep trust he was held in by the President and his record of negotiating political obstacles to get tasks completed. On a formal basis, however, the President justified the appointment as intrinsic to the general reconversion effort.

Steelman's effort was envisioned as a broader effort to catalogue and evaluate both government and nongovernmental research and propose recommendations for future federal involvement with science. Truman asked for an accounting of infrastructure, manpower, programmatic potential and deficiencies, and an overview of the future of science in the "national interest." To assist Steelman, Truman appointed a Scientific Research Board consisting of the Secretaries of Agriculture, Commerce, Interior, War, and the Navy, along with the directors of the OSRD (Bush), the FCC, the TVA, and the National Advisory Committee on Aeronautics (which would morph ten years later into NASA). Notably, no nongovernmental scientists (and nearly no governmental ones) were included on the Scientific Research Board – a point which would later undermine the persuasiveness of Steelman's conclusions and proposals.[14] Truman also singled out training scientific personnel as an area of particular concern.[15]

Steelman, a government administrator by both profession and inclination, viewed the scientific enterprise differently than did Bush. He was less concerned with the actual creative process of science and more interested in the ways in which scientific efforts were coordinated, reconciled, and supported. How did universities actually fund support staff, for example, and how was this effort charged back to the lead investigator? How were the fixed costs of laboratory construction applied to individual projects, and could amorphous indirect charges be fairly placed on a funder's ledger? And, how did a university go about evaluating the quality of the research done under its mantle? This last question was of particular concern, as Steelman pushed back against Bush's recommendation to devolve oversight to the universities and to the individual labs. The Manhattan Project had succeeded brilliantly under the maverick leadership of Oppenheimer, but it had been closely overseen by General Groves and ultimately accountable to executive military command.[16]

"Science and Public Policy," usually referred to as the Steelman Report, was delivered to the President in October 1947. Many of its recommendations were similar to those proposed by Bush: recapitalizing the basic research effort in the face of a ravaged European scientific infrastructure; providing substantial federal support for independent scientists through a competitive grant-making system; substantially increasing funds to train both undergraduate and graduate students in the basic sciences; and investing in European science to reenergize collaboration across borders.[17] Different from Bush, Steelman inserted specific spending goals, notably recommending that 1 percent of the nation's GDP go toward scientific research within a decade and that the biomedical research budget be quadrupled and the clinical research budget tripled. (At the time, the nation's total budget for both basic and applied research was just about $1.1 billion, not including spending on atomic research funded nearly entirely by the AEC.)

Steelman's recommendations for scientific personnel largely paralleled Bush's. At the time that the report was delivered, the nation claimed just about 137,000 working scientists, of which just over 25,000 held doctorates. Without context, it was difficult to judge the adequacy of this workforce. Steelman's concern lay not so much in raw numbers of scientists but rather with the fact that growth in scientific and technical personnel during the war – about 80 percent – had greatly lagged growth in the nation's research spending, which had grown by about 335 percent during that same period. (Growth in the number of doctorally trained scientists lagged further still.)[18] And even that number failed to reveal the fundamental weakness in the training pipeline, where classes had grown larger during the war as teachers and professors were drafted into wartime service. Simply restoring prewar faculty–student ratios at colleges and universities would require the immediate appointment of 15,000 additional instructors.

Like Bush, Steelman envisioned substantial investment in basic research (which was less than 10% of the overall research budget at the time) through a national science foundation, a new iteration of the research foundation which Congress had failed to create. Such investment would be built on individual and institutional grants to private and state universities, along with construction grants for laboratories, accelerators, and large scientific equipment. At the same time, the federal government should continue its own substantial research effort in government labs, of which 80 percent was funded by the Army and the Navy. Disagreements over the optimal balance between civilian-controlled primary research and (largely) military-controlled applied research would roil the federal science effort moving forward. To add urgency to his proposals, Steelman asserted that in World War II the laboratory had become "the first line of defense and the scientist, the indispensable warrior."[19] This would likely be true moving forward.

Both Steelman and Bush framed their proposals with concerns over the tendency of the military to dominate research. Research in weapons, guidance systems, ordinance, and propulsion was politically attractive – such programs generated jobs in industrial labs and plants around the country and produced armaments ripe for political endorsement. Pictures of the atom bombs exploding over Hiroshima and Nagasaki, horrific as they were, provided political cover for members of Congress and the administration, which had endorsed and funded the effort. Theoretical physics and pure mathematics, by contrast, could hardly compete.

Like Bush, Steelman warned of the "dangerously low levels of our stockpile of fundamental knowledge." He warned of some areas in which engineering could no longer progress short of new fundamental discoveries, such as in supersonic flight.[20] By one measure, the nation's efforts in basic research had fallen dramatically during the war; university-based researchers dropped from 49 percent of all scientists in 1930 to 36 percent in 1947, and that was including the many university-based researchers working on Army contracts such as those as the MED and at MIT's Rad Lab. While wartime exigency had dictated a substantial turn to applied military research, future innovation would depend on reversion to a more fundamentalist orientation.

Nearly all of this growth would need to come from federal funds. Few universities, even the wealthiest, had the wherewithal to greatly augment their faculty or to redirect faculty effort from teaching to research. Tuition revenue drove university budgets, even for those universities possessing sizable endowments, and students could not tolerate inflated tuition bills without concomitant commitment to teaching. Were the nation's primary research effort to grow substantially, it would require at least an additional $250 million in increased annual federal spending by 1957 – a 200 percent increase.[21] The government was perhaps the worst culprit, skimming off academic researchers with inflated government salaries for work of marginal value. "The agencies are competing for scientists among themselves in a manner that will prevent any rational distribution of manpower," wrote Steelman to Truman in a follow-up memo.[22]

Oddly, Vannevar Bush was ill-disposed to endorse the report and its recommendations. While he agreed with the general emphasis on federally funded basic research, he felt that Steelman's report had failed to endorse a substantial-enough permanent bureaucracy within the executive branch to effectively guide and coordinate an enlarged federal research effort. Bush feared that Steelman's proposed national science foundation would be a sort of federal cash machine, designed to dispense funds to pet projects of a sitting President rather than direct stable streams of money toward priorities as determined by a board of scientific experts who coordinated efforts with the White House through a new scientific liaison.[23] Bush, the consummate technocrat, agreed with the sentiment of the Steelman Report, but not with the proposed mechanism (however vaguely defined). He wanted to see congressional review of grants, regular rotation of board members, close coordination with the executive branch, and a broad scientific base of expertise for panel membership. "The giving of money to universities and students is a new and important departure which requires a different type of organization," he wrote.[24]

Others were less critical. L. C. Dunn, a zoologist at Columbia University, rued the prewar funding dance of "periodic begging from donors" which robbed scientists of the steady and committed income stream necessary to undertake long-term research projects.[25] And Lee Dubridge, the president of Caltech and former head of MIT's Rad Lab, endorsed the vision more broadly, writing,

> When science is allowed to exist merely from the crumbs that fall from the table of a weapons development program, then science is headed into the stifling atmosphere of "mobilized secrecy" and it is surely doomed – even though the crumbs themselves should provided more than adequate nourishment.[26]

The Soviet Threat

Debates about the direction of science in the United States in the years after World War II were deeply influenced by fears of the Soviet Union. Although a wartime ally and a critical component of the victory over the axis powers, the Soviet Union had shown itself to be aggressively expansionist and ideologically driven. American planners considered Soviet goals and capabilities in nearly every strategy they devised with an eye toward building alliances to counter Soviet expansion. Nowhere was this truer than in postwar science planning.

American planners were prone to exaggerate Soviet designs and capacities, but a motif of ascribed ideological zealotry ran through analytical memoranda. A security analysis of the Soviet threat, delivered to

President Truman in 1950, described the nation as being "animated by a new fanatic faith" seeking to "impose its absolute authority over the rest of the world."[27] Expansion of Soviet influence and troops into Eastern Europe and Central Asia posed an immediate threat to nations allied with the United States and Western Europe. Puppet governments, or heavily Soviet-dependent governments in Bulgaria, Romania, Poland, Hungry, and Czechoslovakia appeared to be the forward edge of an expanding Soviet empire while a spreading sphere of Soviet influence in the Middle East, sub-Saharan Africa, and Southeast Asia appeared to concerned Americans (and western Europeans) a harbinger of hegemony. President Truman's task force reported that the Communist Party of the Soviet Union was an "apparatus designed to impose an ideological uniformity on its people" which would act abroad as an "instrument of propaganda, subversion, and espionage," born of an "utterly amoral and opportunistic conduct."[28]

In the late 1940s, the Soviet Union hardly posed a threat to American economic hegemony. While industrial data coming out of the USSR was deeply unreliable, the best American estimates of the Soviet Union's industrial strength showed the United States to be far ahead along nearly all measurable axes. The United States in 1949 produced four times as much steel as the Soviet Union, five times the aluminum, seven times the electric power, and nearly ten times the crude oil. In agriculture, the United States regularly produced such substantial surpluses as to require removing land from cultivation to stabilize prices, while the USSR regularly purchased grain and livestock abroad to meet the basic nutritional needs of its population. Economically, at least, the United States had little to fear from its rival.

But in science, the difference was hardly so stark. While the United States had been first to radar and to the atomic bomb, the Soviets supported a first-class cohort of physicists, chemists, and mathematicians. While their engineering and manufacturing prowess could not compete with American (or British or German) talents in the years after World War II, this was more a reflection of educational emphasis than raw intellectual might. And the nation's military capacity, both in raw troop numbers and in military–industrial capacity, actually exceeded American capacity in the years of postwar demobilization. Truman's military advisors warned him of the Soviets' ability to overrun much of Europe in the absence of American troop support, to drive to the Middle East, to launch air attacks on the United Kingdom, to sever communications lines between the United States and Western Europe, and to consolidate gains in the Middle East and in Central and Southeast Asia.[29] And America's reserve weapon, the atomic bomb, could hardly alter the scenario initially, given the nation's shortage of fissile material and the growing ability of the Soviet air force to hit manufacturing plants and launch sites. The ever-rational Vannevar Bush, in assessing the Soviet threat in those years, spoke of the "hordes of Russia" who could only be countered by superior armaments born of fundamental scientific and engineering break-throughs.[30] He wrote, nearly hysterically:

> It is impossible to present the extent of this threat in specific detail within the scope of information which is public. Enough has been released to indicate the outlines, and make clear the fact that any action we might take to defend ourselves cuts horizontally through the entire governmental structure. If rinderpest or anthrax were used against our herds, or rusts against our crops, if our water supplies were tampered with, many agencies would become involved at once. If Russia decided to tamper with our currency, or our means of international exchange, a new set are included.[31]

No threat loomed as darkly as that of a Soviet atomic bomb. The great question in 1946 was how far off the Soviets were in developing one, and nobody seemed to know for sure. The original work on nuclear fission and radioactive decay, while European, had emanated from Germany, the United Kingdom, France, and Scandinavia. Russian physicists had produced few substantial breakthroughs or insights into nuclear physics either before or during World War II. Substantial numbers of highly skilled European physicists had fled the continent during the 1930s, leaving some Americans to suspect that the European nuclear research effort generally, and the Russian effort specifically, was moribund or at the very least substantially behind the American effort.

Certainly, Soviet physicists understood the principle of nuclear fission and the mechanics of the U^{235} separation processes. In fact, Americans admitted that in the realms of pure math and theoretical physics, the Soviets were equal to their American counterparts if not actually superior. But it was in translating the fundamentals of nuclear theory into bomb design and production that Americans seemed to have the edge. President Conant, of Harvard, noted that the task before the Russians was applied research and specialized construction – both somewhat lacking as exhibited by that nation's archaic power grid and poor roads. Even if the Russians could produce a bomb within five years, they would lack the capacity to purify fissile material in large quantities for some time to come.[32]

Others were less sanguine. The Soviet Union's totalitarian system had demonstrated in the past that it could mobilize the human and financial capital of the nation to achieve certain national aims in ways that would have been politically intolerable in a democratic state. Bush warned Truman in 1945 that the Russians were no more than five years away from having a nuclear arsenal if they were willing to reduce their standard of living and mobilize their national scientific and engineering resources.[33] And the President's own military advisors offered a timeline for a Russian nuclear stockpile: 10–20 weapons by 1950, 25–45 by 1951, 45–90 by 1952, and over 200 by 1954.[34] Mostly, everybody was confident that the Soviets would develop nuclear arms capability; the questions that troubled American planners were how quickly and in what volume. Countering the threat required not so much thwarting Soviet progress as outproducing them through leveraging the substantial infrastructure that the United States already possessed in its MED sites. The President's National Security Council suggested to him in 1949 that the ultimate deterrent to Soviet force lay not in an American nuclear monopoly but rather in a superior stockpile and production rate.[35]

Underpinning concerns over the Soviet atomic effort was a recognition of Soviet scientific facility generally. The Russians had been highly successful in developing a system of elite and competitive high schools and universities which channeled the brightest and most capable students toward research in service to the state. While talented American youth could choose their calling and schooling from a young age, their Soviet counterparts were seen, to some degree, as assets of the state, whose intellects were to be mobilized to the common good. And the Soviets were rapidly increasing their investment in scientific training. In 1947 alone, for example, the Soviets increased their government outlay on research by a third and initiated a program to train 140,000 new scientists and engineers each year.[36] Karl Compton, the president of MIT, emphasized the critical nature of research in any future war with the Soviets, more so even than in World War II when the United States had been able to leverage the strength of its allies. "Russian advantage in manpower would have to be overcome by a very high degree of technical superiority," Compton wrote to the chairman of the Navy Board in 1948.[37] Bush concurred, pointing out that in a future war, the strategic advantage would lie with nuclear scientists, bomber pilots, and electronic technicians rather than in the raw numerical superiority that had won past wars.[38] He wrote, "The contest of total war is . . . in the development and application of scientific weapons to bring total devastation to enemy territory. . . . It is the evolution of weapons which controls, which is the determinant."[39]

The Office of Naval Research

Although both Bush and Steelman had strongly recommended the creation of a national science foundation to support basic research with federal grants, in fact any such foundation would be overshadowed by the immense scale of defense-related research. By 1947, the Army and the Navy together were funding over 40 percent of the nation's basic and applied research ($500 million of $1.2 billion), with another 35 percent being funded by industry – often on the basis of promised military contracts. Due to the Soviet threat, few military planners thought that this commitment should diminish. Steelman warned that a reduction in military research would be "irresponsible" in the present environment and predicted that the military research budget would continue to grow over the following decade, with little likelihood that the civilian research sector would displace this effort.[40]

The concern, as had been intimated by both Bush and Steelman, was that military research would tend to push scientists toward work with direct applications, ultimately impoverishing basic research and theoretical innovation. Could a contracting system overseen by a career Army or Navy officer grant adequate independence to a civilian scientist to simply pursue work for the sake of innovation? The idea seemed at odds with the military gestalt – one of problem-solving under pressure. Indeed, no less a figure than Dwight Eisenhower, at the time dividing his time between the presidency of Columbia University and consulting with the War Department, warned that future military competitiveness must rest on a cooperative relationship between civilian science and military planning. Fearing overreliance on the military contracting mechanisms, he urged military planners to turn to university researchers in their search for talent, who must be given the "greatest possible freedom to carry out their research." In such relationships, military dictates must be "held to a minimum," and the research relationships must be insulated from ordinary procurement and purchase operations.[41]

Despite these concerns of an over-militarized research enterprise, the OSRD after the war morphed into the Joint Research and Development Committee, which rapidly gave rise to a new Research and Development Board of the Army and Navy.[42] In 1947, partially to compensate for the failure to create a new national science foundation, Congress created a new Research and Development Board (RDB) as part of the National Security Act which combined the War and Navy Departments into a newly unified DoD. The RDB became a critically important mechanism over the following decade to coordinate research between the services, focus priorities, and offer counsel to the Secretary of Defense, the newly created Joint Chiefs of Staff, and the House and Senate Armed Services Committees.

Although the nature of military research necessitated engagement in civilian science and in problems not clearly applicable to military need, the RDB was unmistakably a creature of the military. While its first two chairmen were Vannevar Bush and Karl Compton, the other six members of the board were all senior military officers – two each from the Army, Air Force, and the Navy. From the onset, it viewed itself not as a proxy for a national science foundation but rather as a unique body aimed at optimizing military research, even as that research required underlying basic explorations. The national science foundation, should it ever be created, would concern itself exclusively with nonmilitary efforts.

But this focus raised the perennial challenge of resolving tension between scientists and military officers. A 1950 report to the RDB, from the Orwellian-sounding Committee on Plans for Mobilizing Science, noted that many of the best scientists – those with the most inquiring minds and innovative miens – would balk at a rigid chain of command. "A nation can lose a war just as easily by putting its top-flight scientists at the wrong work as by sending its ships, planes, or troops to the wrong place," the committee warned. "Many of the most creative scientists will not volunteer for work, or work effectively, unless afforded freedom to a degree impractical within military organizations."[43] At the same time, an RDB which was incapable of directing ("coordinating") a research agenda would add little to military preparedness – the *sine qua non* of the board. Within a few years of the board's establishment, complaints surfaced that its recommendations were arbitrary, that (to quote an internal investigation), "we make decisions without knowing their effect, that is, without thorough analysis."[44] Even a coherent *philosophy* of control seemed elusive: was the goal a focused research enterprise or rather some sort of inspired chaos? One internal planning memorandum at the time made the case that the best government laboratories were defined by "a single clear objective" rather than being "merely given an area of work" – the latter being the essence of a grant-funded academic investigation. What was the answer?[45]

Until a national research foundation could be established, the answer lay in the Office of Naval Research (ONR), created by an act of Congress in 1946. ONR was an administrative stepchild of the ever-enlarging Naval Research Laboratory (NRL), created shortly after World War I in Washington DC. NRL had largely focused on ordnance and maritime technology and navigation, leading the way on long-range radar in the 1920s and devoting itself to anti-submarine warfare during World War II. With a staff built of almost exclusively civilian scientists and engineers, NRL provided a template for applied civilian research in service to military concerns.

ONR grew rapidly after its creation. Building on its intramural research program, it created a large contracting program with civilian scientists based in both industrial concerns and at universities and committed itself to reimbursing the full cost of research including overhead costs and non-salary personnel requirements as well as general research expenses. In its first year it spent $10 million on academic research, which increased to $25 million a year later. *Newsweek* called ONR the "Santa Claus of basic physical science," having quintupled the federal government's peacetime contribution to basic research in the space of just two years.[46] By 1949, defense contracting coupled with the newly created AEC accounted for 96 percent of all federally funded research in the physical sciences in the nation's universities.[47]

ONR was unusual in its commitment to training and research grants beyond its military purview and to buttressing academic freedom. Under the leadership of Alan Waterman, who had moved from his wartime position as the director of field services for the OSRD, ONR supported graduate students in the engineering and physical sciences and underwrote academic research in a manner not usually associated with military contracting. Waterman was particularly concerned with fostering a culture of scientific transparency, noting that high-quality science was incompatible with restrictions on the "free flow of information." Moreover, Waterman committed over three-fourths of all ONR contracts in the early years to university-based scientists, whom he felt were best positioned to conduct innovative research.[48] Isidor Rabi, a lead scientist at MIT's Rad Lab during the war, considered

Waterman's policies at ONR critical to American scientific competitiveness in the decade after World War II. Rabi reflected:

> The general health of our research universities, and elsewhere, I think, is largely due to the efforts of the ONR. . . . We really would have been way behind the Russians . . . if ONR had not done it . . . I think we would have been in much worse shape.[49]

During the five-year delay in creating the National Science Foundation (NSF), ONR became critical to the nation's basic scientific enterprise, devoting over 90 percent of its university-targeted grants to projects without clear naval applications. Captain R. D. Conrad, the director of ONR's planning division, told an audience in 1946, "It will be a long time before a National Science Foundation can be enacted into law. . . . In this critical interval, the armed services are providing the support so essential for science."[50] In the early 1950s, ONR funded 75 percent of industrial research in aerospace and electronics. Historian Audra Wolfe writes that in the years before the founding of the NSF (in 1950), ONR had become the *de facto* Office of National Research.[51] Isidor Rabi credited ONR with maintaining university-based research during those years and maintaining a technological edge over the Soviets. "They cannot be given too much credit for that accomplishment," he wrote.[52] And Harold Bowen, the chief of the agency, concluded upon his retirement, "You could move the Office of Naval Research to another building, put up a new sign on the door reading 'National Science Foundation', and you would have the nucleus of such an agency."[53]

Funding from ONR, while modest within the context of the size of the higher education sector, began to influence decision-making in the nation's research universities. The two-dozen institutions which conducted the majority of the nation's basic research were initially skeptical of the flow of federal funds, leery of military control and of the long-term stability of the funding stream. Rapidly, however, academic leaders shifted their hiring and staffing plans to build more research-intensive faculties, which would rely on federal grants rather than on tuition. By 1949, ONR was funding 1,131 projects at 200 universities using its own funds as well as managing funds made available by the newly created AEC – constituting nearly 40 percent of the nation's work in basic science and nearly 80 percent in the physical sciences.[54] As funding became available, applications for the funding increased rapidly; by 1950, the agency was receiving four times as many applications as it could fund. Funds went out for research in pure math, cosmic rays, meteors, cyclotrons, synchrotrons, biopsychology, neurological impulses, and white dwarf stars. ONR funded early efforts at digital computing, statistical analysis, and entomology. While the bulk of the agency's grants, 75 percent, went to the physical sciences, only 10 percent of all grants went to projects that were within the traditional purview of naval science. One spokesman indicated that ONR cleaved to its mission of funding projects deemed "useful" to the Navy but that the agency applied a "liberal" reading to the standard.[55]

As research universities began to rely on ONR funding, they became increasingly invested in the stability of that funding and in the politics underlying the federal budget process. Within two years, Waterman's office at ONR was receiving multiple letters from deans, provosts, and research directors to inquire about the state of the next year's funding and the long-term viability of ONR's grants program. In 1948, Lee DuBridge, president of the California Institute of Technology, wrote to Waterman expressing his concern over proposed cuts in the next year's ONR budget. He queried,

> I recall that one was planning to ask for a budget large enough to be able to extend its contracts for a three-year period. . . . I wonder if there has been any action on the part of the Bureau of the Budget, whether there has been a cut, and if there has been a cut, whether it affects the proposal to lengthen the contract periods, or whether it affects also the annual operating rate proposed for next year.[56]

Waterman assured DuBridge that his budget was safe.[57] The advisory board worked to calm worried deans or at least keep them updated. William McCann, a member of ONR's research advisory committee, wrote to the dean of Harvard Medical School: "Unless we rescue the research program, it will be seriously cut and about two hundred out of three hundred contracts will have to be cancelled."[58]

ONR in the late 1940s was an odd creature. Despite the warnings of Bush and Steelman, the nation's research agenda, particularly in the physical sciences, had fallen to the Navy rather than to civilian leadership. The scope of the naval budget, Cold War pressures on federal priorities, and the military orientation of ONR's predecessor organization – the OSRD – all eased the way for the Navy to emerge as the federal government's leader in funding science. Yet despite many planners' fears, ONR's approach was ecumenical. Committed to scientific openness and to basic research, it established competitive grant-making procedures for university-based science which largely fulfilled Bush's vision for postwar American science as laid out in his 1945 manifesto. Universities, initially leery of partnering with the Navy in the aftermath of wartime exigency, quickly pivoted. The Navy was proving to be an excellent friend of American research universities, of high-level science, and of scientific training. Even when the NSF eventually came into being, ONR would maintain its position as a critical agency in American research.

Notes

1 See "Minutes of the President's Scientific Research Board," January 2, 1947, HST, SRB, 3: "Minutes of 1st Meeting." Also, FDR to VB, November 17, 1944, DB, FA 965, 38: OSRD Bush, Van (1942–46).

2 See Bruce Smith, *American Science Policy Since World War II*, Brookings, 1990, pp. 40–44.
3 Bush, *Science, The Endless Frontier: A Report to the President*, GPO, July 1945, p. 72.
4 "Report of the Subcommittee on the Bush Report," HST, SRB, 2: Science Legislation Pubs.
5 "Statement of Vannevar Bush at the Joint Hearings Held by Subcommittees of the Senate Committee on Commerce and the Senate Committee on Military Affairs," October 15, 1945, p. 3.
6 Bush was careful to include young women in his remarks regarding the undertraining of the nation's young. "We should take affirmative steps to make certain that young men and women with a capacity for scientific achievement, but lacking in funds, are enabled to receive the required scientific training." Ibid., p. 5.
7 Ibid.
8 Ibid., p. 6.
9 "Statement of Detlev Bronk at the Joint Hearings Held by Subcommittees of the Senate Committee on Commerce and the Senate Committee on Military Affairs," October 15, 1945, p. 8.
10 Ibid.
11 Vannevar Bush, "Report to the President on the Office of Research and Development in the War," October 19, 1945, VB, 3091: Report of the President.
12 "For Immediate Action," February 13, 1946, DB, FA 965, 37: NRC Post-War Legislation (1945–46).
13 Steelman is often called the first White House Chief of Staff, although he never officially held that title.
14 "E.O. 9791: Statement by the President," October 17, 1946, HST, SRB, 2: Science Legislation Publications. A proposal to create an advisory board, which might have included such scientific luminaries as Oppenheimer, Einstein, Arthur Compton, and Irving Langmuir, was shelved.
15 "Statement by the President," October 17, 1946, VB, 2144.
16 From an early draft of the foreword to the final Steelman report, HST, SRB, 2: PSRB.
17 Interestingly, Steelman also highlighted the loss in access to Soviet science in reconfigured postwar national alignments. He wrote, "In other [parts of the world], an iron curtain has been drawn around the work of the scientists. That free exchange of ideas which formerly permitted us to import to meet our needs no longer prevails." John Steelman, *Science and Public Policy*, GPO, August 27, 1947.
18 Ibid., pp. 15–16.
19 Ibid., p. 2.
20 Ibid., p. 4.
21 Ibid., p. 29.
22 Steelman to Truman, September 23, 1946, HST, SRB, 2: PSRB.
23 Bush's view was laid out in a working paper authored by one of his senior staff members, Lyman Chalkley, who called for a permanent presidential science policy committee and the appointment of a special assistant to the President for science policy. See Chalkely, "A Science Policy Committee," May 21, 1947, HST, SRB, 3:" Comments on PSRB Reports," and Chalkely, "A White House Office of Scientific Liaison," June 4, 1947, HST, SRB, 6: Drafts – PSRB.
24 Bush to James Forrestal, September 10, 1947, VB, 2144.
25 L. C. Dunn, "Organization and Support of Science in the United States," *Science*, 102:2657, 1945, p. 548.
26 Quoted in Dan Kevles, "Cold War and Hot Physics," *HSPBS*, 20:2, 1990, p. 241.
27 "Report to the President," April 7, 1950, HST, PSF, 174: Atomic Energy Advisory Committee, p. 1.

28 Ibid., pp. 3, 4.
29 Ibid., pp. 9–10.
30 Bush to Conant, January 18, 1951, VB, 614.
31 Ibid., p. 4.
32 Conant to Clark, October 8, 1945, VB, James Conant.
33 Bush to Truman, September 25, 1945, HST, PSF, 174: Atomic Bombs, Cabinet, James Byrnes.
34 "Report to the President," April 7, 1950, HST, PSF, 174: Atomic Energy Advisory Committee, p. 12.
35 "Report to the President by the Special Committee of the National Security Council on the Proposed Acceleration of the Atomic Energy Program," October 10, 1949, HST, SRF, 11A:5.
36 Steelman, *Science and Public Policy*, p. 4.
37 Compton to McMorris, April 14, 1948, MIT, 156: Department of the Navy.
38 Untitled strategy memo, 1945, VB, 3091.
39 Ibid., p. 13.
40 Steelman, *Science and Public Policy*, p. 20.
41 Eisenhower, "Scientific and Technological Resources as Military Assets," April 30, 1946, MIT, 195: SAC.
42 During the war, there was an effort to create a Research Board for National Security to coordinate civilian and military research after the war, under the influence of Merle Tuve, a physicist at the Carnegie Institution. Ultimately, the defense-based RDB, along with the creation of the Office of Naval Research, undermined the case for such a board. See Daniel Kevles, "Scientists, the Military, and the Control of Postwar Defense Research: The Case of the Research Board for National Security, 1944–46," *Technology and Culture*, 16:1, January 1975. The original org chart for the JRDC can be found at VB, April 6, 1946, 1403.
43 "Report of the Committee on Plans for Mobilizing Science," June 26, 1950, RDB report series no. 39, p. 5.
44 Arleigh Burke, "Suggestions for Improvement in the Effectiveness of the RDB," March 15, 1950, AW, 32: PSAC.
45 "An Appraisal of Some Indicated Needs of Defense Research," December 3, 1951, MIT, 194: SAC.
46 Quoted in Greg Pascal Zachary, *Endless Frontier*, MIT Press, 1999, p. 329.
47 Audra Wolfe, *Competing with the Soviets*, Johns Hopkins University Press, 2013, p. 27.
48 Alan Waterman and Robert Conrad, comments on Louis Ridenour "Should the Scientist Resist Military Intrusion?" *American Scholar*, Spring, 1947, found in AW, 31: ONR.
49 Quoted in "Contributions of Dr. Alan T. Waterman to Naval Research and Development," AW, 32: PSAC.
50 From a speech in Terre Haute, Indiana, October 18, 1946, quoted in Philip Powers, "A National Science Foundation?" *Science*, 104:2713, December 27, 1946, p. 618.
51 Wolfe, *Competing with the Soviets*, p. 26.
52 Quoted in "Contributions of Dr. Alan T. Waterman to Naval Research and Development," AW, 32: PSAC, p. 2.
53 Quoted in Greg Pascal Zachary, *Endless Frontier*, p. 329.
54 John Pfeiffer, "The Office of Naval Research," *Scientific American*, 180:2, February 1949.
55 Ibid., p. 14.
56 DuBridge to Waterman, November 26, 1948, AW, 31: ONR.
57 Waterman to DuBridge, December 2, 1948, AW, 32: OSRD.
58 William McCann to Sidney Burwell, January 20, 1947, VB, 3091.

4 The Atomic Energy Commission

The Atomic Energy Act

In the aftermath of Japan's surrender, the nation looked forward to substantially shrinking the military. Troops could be sent home, warships decommissioned, wartime bases shuttered. Rising tensions with the Soviets precluded the sort of comprehensive drawdown of previous wars, however. A power vacuum in Japan had left the United States Army as the principal governing force in that country, with General Douglas MacArthur acting as the head of government, even as an American-led delegation drafted a new constitution for a demilitarized democratic state. In Germany, four independently governed quadrants were governed by occupational forces of the United States, the United Kingdom, France, and the Soviet Union, with the first three quickly confederating into the new democratic state of West Germany (officially the Federal Republic of Germany). By 1949, West Germany achieved some degree of independence from its military governors when Konrad Adenauer, the leader of the Christian Democrats, took over as the rump-nation's first prime minister, committed to policies of de-Nazification, democracy, economic growth, and pro-Western alliances. East Germany (the German Democratic Republic) achieved far less independence in the immediate postwar years, as the Russian-influenced Socialist Unity Party closed borders, halted emigration to the West, and governed under the watchful eyes of nearly 1,000,000 Soviet troops. The fraught and complex situation forced the United States to carefully consider which of its military assets could be safely decommissioned and which would need to stay active.

The atomic bomb posed a unique challenge to demobilization. The United States' massive wartime investment in developing the MED had left the nation as the sole possessor of atomic infrastructure and capability. Leaders in both the scientific and political communities debated the disposition of this infrastructure, with more progressive-minded scientists wanting to turn the whole thing over to an international governing committee with monopoly control over all fissionable material and machinery and hawkish members of Congress wanting to grant a perpetual monopoly to the US military. (Few thought it wise to grant the private sector freedom to develop

DOI: 10.4324/9781003363897-4

atomic technology in pursuit of profit.) Several of the most leftist scientific groups, drawing their ranks from veterans of Los Alamos, Oak Ridge, and the Chicago Met Lab, coalesced into the Federation of Atomic Scientists in 1946, with the stated aim of "furthering world peace and the general welfare of mankind."[1] Dovish members of Congress concurred, expressing a fear of "losing the peace" which had only recently been gained. Senator Joseph Ball, of Minnesota, a particularly vocal spokesman for this faction, warned of the dangers of a toothless United Nations in the face of a proposed nuclear arms race. "Far from strengthening our collective security organization," he lectured in Fall, 1945,

> [W]e are already embarking on a secret, nationalistic armaments race in the invention and production of even more destructive weapons with ever greater ranges. . . . the climax of every such race in the past has been an explosion into war.[2]

A consensus rapidly evolved that atomic assets needed to remain within the purview of the federal government, albeit under a modified governance structure.[3] An initial bill introduced into Congress by Representative Andrew May and Senator Edwin Johnson that Fall placed the preponderance of authority and discretion with the military, creating an oversight panel dominated by career officers with all affiliates held to high levels of secrecy and with an emphasis on directing nuclear research toward military applications. The bill was opposed by prominent atomic scientists – notably Leo Szilard and Harold Urey of the University of Chicago – who were concerned that the permanent administrator and deputy administrator of an eventual commission would be drawn from senior military ranks and thus beholden to the priorities of the War Department. Other scientists, as well as more liberal members of Congress, opposed it on the grounds that it would squash the free exchange of ideas and discoveries while underfunding research toward eventual civilian applications of nuclear power.

The scientists who had opposed the May-Johnson bill predicated their opposition on the senselessness of military monopoly on atomic knowledge. Represented by a variety of associations – Atomic Scientists of Chicago, Association of Oak Ridge Scientists, Association of Manhattan Project Scientists – they argued through 1945 and into 1946 that the atomic bomb was not based on any great scientific secret; that the principles of nuclear fission were widely understood; and that even the technology behind uranium enrichment could rapidly be replicated by any nation with advanced metallurgical industries. Given that other nations would inevitably develop their own atomic bombs over time, the more constructive response, indeed the only realistic response, was to create an international commission to regulate and control atomic research and technology and uranium enrichment and distribution. Historian Jessica Wang enumerates the basic components of the activist–scientists' platform: "1. The A-Bomb is no secret. 2. We cannot long have a monopoly of its manufacture. 3. International control is the only solution."[4]

Scientists and political leaders debated several points touched by atomic regulation, with no obvious answer. Could the Soviets be appeased by sharing nuclear secrets, as Ball suggested, or would they simply appropriate any shared information toward their own expansionist aims? Should the United States share its insights, reactor technology, and enriched materials with a growing cohort of European allies, or should control of fission remain the exclusive province of the United States? More than a few scientists scoffed at the very idea of nuclear secrets, suggesting that the principles behind the enrichment process and the know-how to actually build a sustainable reactor pile were already widely known among advanced nations and that the challenge was simply one of engineering and machine tooling. "The discoveries which enabled us to produce it [the atomic bomb] can be duplicated by any industrial nation in from three to ten years," predicted Ball.[5]

An alternative bill, cosponsored by Helen Gahagan Douglas in the House of Representatives and Brien McMahon in the Senate, created a smaller but stronger civilian committee (whose members would serve full time rather than part time), which would grant greater freedom to researchers. The McMahon bill pledged fealty to any international atomic arms pacts which might be signed in the future and to the free exchange of ideas, discoveries, and data. While the bill left the government's atomic monopoly intact, it freed atomic research from most military oversight and inaugurated a substantial academic nuclear research enterprise. The modified bill passed both houses of Congress the following summer and was signed into law by President Truman in August 1946. The newly created Atomic Energy Commission (AEC) came into existence on January 1, 1947 – the successor agency to the MED, and the inheritor of its properties, plants, bureaucracy, and secrets.[6]

Authority in the new AEC flowed from five full-time civilian commissioners who were paid quite generously – $15,000 per annum – with an additional $2,500 stipend for the chairman. The members, chosen for their experience in senior decision-making roles in business and academe and for their scientific and engineering expertise, would not hold political office or positions in the armed forces or the civil service during their term of service. David Lilienthal, the first chairman, exemplified the background legislators had envisioned, with his many years of service on the Wisconsin Public Utilities Commission, as a lawyer in private practice, as the chairman of the Tennessee Valley Authority, and as a coauthor of the Acheson-Lilienthal Report. Administrative authority was invested in a presidentially appointed general manager who answered to the commission. Under the general manager were four division chiefs heading up the operational components of the commission: research, engineering, production, and military application. The only stipulation for any of these roles was that the division director for military affairs should be an active duty member of the armed forces.

Congress structured the AEC in such a manner that commissioners had multiple points of input and multiple masters. The General Advisory Committee (GAC), composed of nine highly accomplished nuclear scientists and engineers, provided the commissioners with scientific expertise. The office of military liaison tied the work of the commission to the concerns of the military. A new bipartisan Joint Committee on Atomic Energy, drawing its members from both houses of Congress, ensured that the commission's work would be consistent with the priorities of the nation as translated through the political process. And eight additional advisory committees provided scientific expertise to the commission on such matters as biology and medicine, exploration and mining, patents, medical safety, and isotope distribution.

On January 1, 1947, the AEC took over the assets, properties, and staffs of the MED, instantly catapulting it to the ranks of a major federal agency. Immediately, the new commission had 55,000 employees and an annual budget of nearly $700 million.[7] The commission now had sole authority over the "big three" MED properties (Los Alamos, Oak Ridge, and Hanford) as well as a dozen smaller labs and operations at Brookhaven, Argonne National Laboratory (outside of Chicago), Berkeley, Los Angeles, Ames (Iowa), and Schenectady (New York). It even owned rights to test atomic bombs on the Eniwetok Atoll in the South Pacific.

The AEC's strongest lever in ensuring a nuclear monopoly for the US government was control of enrichment and distribution of fissionable materials. The great majority of investment, infrastructure, and personnel during the years of the Manhattan Project had been devoted to separating out the minute amounts of U^{235} from uranium ore and cooking plutonium in the breeder reactor at Hanford. Both tasks, while simple in theory, were enormously difficult in practice. To separate U^{235}, engineers had to gasify uranium-hexane and then spin it at tremendous speeds (12,000 revolutions per minute) over many hours to gradually build up concentrations of the rarer isotope. Producing plutonium was even more difficult, as the newly created plutonium (a product of neutron capture in an atomic reactor by ordinary U^{238}) was mixed in with uranium ore. The whole mess needed to be gasified and the plutonium separated from numerous other elements and isotopes in the slag using both chemical and physical processes. Among other challenges associated with the process was the high volume of radioactive waste created during the process (which needed to be disposed of underground) and the extraordinary amount of cooling water required to ensure that the pile did not melt down.[8] At its peak, the Hanford reactor was drawing off 75,000 gallons of water from the Columbia River *each minute.*

In creating the AEC, Congress was most explicit about control of fissionable materials, whose production by any private person or firm was simply illegal. The great concern of Congress and the military in 1946, however, was not so much private American firms developing their own enrichment processes, as foreign nations replicating the work of the MED. Fortunately, the footprint for an enrichment and separation plant was immense. Building Oak Ridge had required 25,000 workers and 33,000 carloads of freight, which in turn required constructing new rail capacity and a small city to house workers and engineers, all in an effort to produce enough enriched uranium to build between one and five bombs per year. Greater production would require even larger infrastructure.[9]

Although Congress was careful to demilitarize the AEC at conception, in fact military concerns were central to the work of the commission from the start. The tremendous fear of a Soviet atomic program drove the scientists on the GAC to spend their time in the first year considering questions of military strength and strategic parity and guessing at Soviet progress and capacity. Lee DuBridge, one of the original members of the GAC, enumerated the major questions facing his group starting in 1947:

> How large a stockpile of fissionable weapons was required to assure national security? When or under what circumstances, if at all, should the vastly more powerful thermo-nuclear weapon [the hydrogen bomb] be developed and produced? How should the responsibility for the development and operation of nuclear power reactors be divided between government and private agencies? To what extend shall nuclear weapons be tested in the atmosphere, underground, or in outer space? What are the "tolerable" limits of atmospheric radioactive contamination? What aspects of the science and technology of nuclear energy must be kept secret, which ones "declassified," and how shall security measures be made both acceptable and effective, without impeding progress?[10]

Already by 1947, the commissioners were faced with substantial challenges in the AEC's military mission. The agency had little weapon's grade material (that is, uranium or plutonium which had been purified to a level necessary for use in bombs – about 20 times the purity level of reactor-grade fuel). None of the material had actually been assembled into bombs. Plutonium production at Hanford had fallen off sharply, nuclear fuses were deteriorating more rapidly than had been expected; and the Hanford reactor appeared to be aging so rapidly that engineers predicted its useful life at no more than 18 months. Nobody could conceive a better way to store the highly toxic radioactive waste being generated at Hanford other than simply burying it in insulated storage tanks. And perhaps most importantly, engineers had yet to figure out how to make plutonium production more efficient. At the time only a very small portion of the U^{238} was being converted to neptunium (and henceforth to plutonium) meaning that most of the fuel – possibly 99 percent – was simply going to waste. With the world's known uranium stocks in short supply, AEC commissioners estimated that Belgian mines, which were the principal supplier to the US atomic effort, would be depleted in two years.[11]

Despite these concerns, commissioners did not devote their attention solely to military matters. The promise of power production – whether for residential use or for specialized transportation solutions in submarines and ships – beckoned. By 1947, British labs had already managed to draw off minute amounts of heat for power production from experimental piles – about 100 kilowatts worth. But meaningful power production would require tweaking the design of nuclear reactors to allow them to run at much hotter temperatures while reliably drawing off the heat to produce steam. Two challenges lay before the commission and its contracting researchers: contamination (any coolant run through a reactor came away dangerously radioactive) and the danger of a meltdown, wherein the reactor core overheated to such an extent as to melt right through the floor of the reactor building and into the ground below. Paper schemes were easy to produce ("I could give you so many drawing of reactors you'd have to haul them away in a truck," explained Walter Zinn of the Argonne Laboratory in 1949) but actually building the experimental piles was expensive and risky and entailed making choices of design, construction material, and cooling systems.[12]

Moreover, the AEC faced difficulty in luring the most creative engineers to its civilian mission. While physicists flocked to AEC-funded research projects, engineers were more hesitant. The challenges to producing civilian applications of atomic energy were vast; the operations were shrouded in secrecy and true authority was retained by federal administrators. The smartest and most ambitious civilian and academic engineers were reluctant to partner with such a forbidding funder, all the more so because of a general (though misguided) belief after the war that many of the permanent scientists within the AEC were mediocre. Government salary caps were part of the problem – senior government scientists earned only a fraction of the salaries of their counterparts in the private sector – but the broader issue was control and secrecy. A pessimistic Karl Compton wrote to his brother Arthur, the chancellor of Washington University:

> While our scientists are eager and enthusiastic about work in nuclear science, we have been able to discover or to stir up relatively little interest in the subject in our engineering groups. . . . The general reaction is that 1) the applications are remote, 2) even if successful the field will be under government control which is not attractive, and 3) there are other attractive problems and possibilities which are closer at hand and which do not involve the uncertainties and the same degree of rigid governmental control.[13]

By mid-1949, the AEC had built one new plutonium reactor at Los Alamos and was nearing completion on a new research reactor at Brookhaven (Long Island). Four additional experimental reactors were in design, notably a more advanced breeder reactor (a "fast-neutron" type rather than the low-energy piles based on the original Fermi-build prototype) and a land-based smaller reactor, which could conceivably be placed aboard a naval ship or submarine in the future. At the same time, the Commission sought a 400,000-acre site in the west to build a new Hanford-style complex where further breeder reactors could be developed with the long-term goal of building an economically viable source for electric power production. In the distant future, AEC commissioners, working with the military liaison and the GAC, envisioned a possible aircraft-based reactor, as well as a "homogenous" reactor in which the nuclear fuel and the graphite substrate (critical to slowing down neutrons to speeds at which they could be captured by U^{235}) would be combined in a cast mixture, thus precluding the need for the expensive and meticulous reactor construction.[14]

Civilian Control

The Atomic Energy Act, passed rather rapidly in 1946 (in legislative terms), rested on the cracked foundation of fundamental disagreement over civilian control. During the war, the MED had been created amid wartime exigency in which military priorities were paramount; no civilian agency seriously challenged the Army's control of the project, nor questioned Roosevelt's decision to exploit the nascent science of nuclear physics to military ends. But even amid the strange and frightening polar realignment of the United States and the Soviet Union after the war, scientists, legislators, and government officials squared off in debating the appropriate roles of the private sector, the military, and civilian authority in controlling atomic energy moving forward. Although almost everybody understood that the military had a powerful (indeed, superlative) interest in the future of atomic energy, the question was whether the military should continue its wartime hegemony. "There is a vast difference," wrote Roosevelt's Secretary of Commerce Henry Wallace, "between entrusting to the Service Departments responsibility for the military applications of atomic energy and vesting them with power over all matters in this field relating to the common defense and security."[15]

Perhaps the most ardent proponent of military control was Vannevar Bush, who had exercised near-total control of all science related to the war effort during his four years heading the OSRD. From his perch at the Carnegie Institution, he voiced near-hysterical warnings about the oncoming "total war," the need to develop weapons which could bring "total devastating to enemy territory," and the compressed timeline under which the next war would be fought. He warned,

> The period of active hostilities in World War III will be too short for much coherent evolution of weapons to proceed. Our strategic and tactical, offensive and defensive, preparations, therefore, must center not in the exercising of troops with the obsolete weapons of World War II, but in the evolution during peace of the new weapons of World War III.[16]

Above all, he told the Senate Special Committee on Atomic Energy in 1945, "For the years immediately before us, the deadly rather than the beneficent power of atomic energy will continue to hold first place in men's minds."[17]

Senator McMahon's effort to wrest military control over atomic energy (in contrast to the failed May-Johnson effort) was, in fact, nearly derailed by the addition of an amendment proposed by Senator Arthur Vandenberg (R., MI) to expand military control over the proposed AEC. Physicists associated with the Federation of American Scientists (FAS) strenuously resisted, suggesting that formulating the new AEC in a fundamentally military posture risked signaling aggressive intentions to the Soviet Union. Eugene Rabinowitch, the University of Chicago physicist who had cofounded the FAS, warned: "Permanent military control of atomic energy in America will signify to the world that America is basing its long-range policies on the assumption that a new war is inevitable, and this will help to make it inevitable."[18] A policy of paranoia could not help but exacerbate tensions and accelerate an arms race while undermining any possible negotiations to ease tension. Maintaining a strict veil of secrecy would only make things worse. In a letter of December 1947, Oppenheimer (with the concurrence of the entire GAC) warned President Truman of the "adverse effects of secrecy, and the inevitable misunderstanding and error which accompany it," and that the "fruits of secrecy are misapprehension, ignorance, and apathy."[19] Historian Wang writes of the scientists' concerns:

> To scientists such as Rabinowitch and [Edward] Condon, the larger implications of internationalism in science led to the conclusion that scientific freedom and international relations were intimate connected. Secrecy and military dominance of atomic energy held dangers far greater than damage to scientific research. Secrecy constituted a mindset that created a false sense of security that defense of the United States could be maintained by the preservation of the atomic monopoly.[20]

Ultimately, the Vandenberg amendment was diluted to an acceptable level to garner FAS support, but the shadow of the amendment tinged the final bill. When the Atomic Energy Act was passed in late 1946, its language included the death penalty for any employee or contractor who communicated classified content to an enemy, drawn from any "document, writing, sketch, photograph, plan model instrument."[21]

Even with legislated civilian control, the AEC faced consistent pressure from the military to erode its franchise. In 1948, for example, Secretary of Defense James Forrestal requested that the stockpile of existing atomic weapons (admittedly quite small at the time) be transferred immediately to the armed forces, along with accompanying fissionable material to be stored for inclusion in future atomic weapons.[22] At the same time, the military applications division of the AEC ballooned into the largest operational division of the agency as it pressed forward on projects of direct interest to the armed forces, such as the Nuclear Energy for the Propulsion of Aircraft (NEPA) project, whose goal was to put a nuclear reactor on board the air force's long-range bombers, a submarine-based reactor, and smaller atomic bombs for attacking military targets with surgical strikes.[23] In a 1948 report, the President's Special Committee on National Security proposed accelerating the program of atomic weapons development and testing to create an adequate stockpile to defend western Europe and to make broader use of atomic weapons in a variety of theaters around the world. The authors counseled:

> As soon as practicable the element of intrinsic scarcity must be eliminated as the predominant consideration of atomic weapon use in war in order to allow the Joint Chiefs of Staff greater flexibility to plan as desirable the employment of atomic bombs for operations where they could be employed more economically than other military measures.[24]

And what of civilian use or even civilian control? There was no support for civilian access to atomic weapons, but civilian engagement in nuclear research toward peaceful ends hardly seemed outlandish. The potential for atomic energy to power electric power plants was obvious from the start; one of Fermi's first calculations upon achieving sustained and controlled fission was the equivalency in freight cars of coal to a pound of fissionable uranium. Oppenheimer and his colleagues on the GAC rued the displacement of civilian power production by military uses of enriched fuel, writing: "If the atomic materials were not required for military purposes they would be available on a scale large enough to be an important fuel for civil power."[25] Bush, hawkish as always, agreed that ultimately ("surely in a reasonable time") atomic research would give rise to electric power production.[26] And economic conservatives and libertarians expressed concern of government monopoly of the new enterprise, in which patent protections were "wiped out", all ores of uranium and thorium required federal licenses to extract, and all contractors would work with the AEC on a nonprofit-plus-cost base (no financial incentive for innovation or efficiency). Lilienthal, who had spent the early part of his career representing private energy companies before moving into the executive role at the TVA, ruminated on the role of private enterprise in the exploitation of nuclear power. "Somehow we must see to it that there will be something in it for industry. At the moment, nobody is quite wise enough to figure out how that will happen."[27]

President Truman was won over by these arguments. With his roots in small business and municipal politics, he was naturally drawn to practical applications of nuclear power, even as he acknowledged the gravity of Cold War concerns facing the nation. In commenting on the AEC's first report of the agency's progress, he emphasized that a free society places "civil authority above military power," and that control of atomic energy belonged in civilian hands. Even in the context of heightened security concerns, Truman argued, the nation must continue to probe nature, expand its understanding of atomic energy, and both "supplement our defenses" while remaining open to "new opportunities for peaceful progress."[28] Military concerns would remain paramount, but civilian goals could not be brushed aside.

Operations

At the outset, the AEC was granted a certain latitude in defining its goals but within the constraints of Cold War military concerns. Building on a generous endowment – the legacy of the MED – the agency's budget soon expanded to over $400 million per year with five principal installations, 20 secondary ones, 50,000 full-time personnel, and many hundreds of contractors and subcontractors. The total land holdings of the commission – some 3,000 square miles – was equal to half of New Jersey. The value of its assets exceeded that of General Motors by a comfortable margin. By 1951, with several new facilities on the line, it was the country's largest consumer of electric power. By 1949, the commission could claim to manage or own the largest and one of the most complex systems of laboratories in the world.[29]

But all was not well. Surveying the state of the AEC's facilities, and cognizant of the critical role which atomic energy would play in an expanding Cold War, the GAC was not initially sanguine. Although in 1947 the United States possessed, by far, the largest and most advanced nuclear research and manufacturing capabilities in the world, already these facilities were showing their age. Oak Ridge, Hanford, and Los Alamos had been hurriedly constructed during the war and were rapidly deteriorating. The nuclear effort, always somewhat decentralized, was poorly focused. The different installations were failing to coordinate their efforts with each other, and each was evolving toward an independent postwar mission. Existing facilities, built to accomplish a narrow set of wartime objectives, were proving inadequate to moving atomic research and development to a next stage. Oppenheimer summed up the commission's challenges:

> Wartime installations and laboratories, which served so well their primary functions of developing atomic weapons for early military use, were in most cases not suited to continue the work as the nature of the technological problems altered, as the transition from wartime to peacetime orientation changed the conditions under which rapid progress might be possible.[30]

No one mission emerged initially to dominate the others. From the start, the AEC had a trifold directive: weapons development, reactor development, and basic atomic-related research in the physical and biological sciences. And while the three were very much interrelated, both the commissioners and the GAC (along with other advisory committees) were leery lest the demands of applied military work (the first goal) overwhelm commitment to the other two. But the latter two goals were hampered by the twin barriers of secrecy and government monopoly, both of which undermined free inquiry and private initiative. Moreover, secrecy threatened to derail scientific careers, as under AEC guidelines work that was "born secret" remained secret, as did all future derivatives of that work. When Kenneth Pitzer became the director of research at the AEC, he faced the challenge of luring high-quality scientific talent to the secretive AEC labs despite the bogeymen of militarism, stifling bureaucracy, and a perpetual veil of confidentiality. *Time* magazine described his challenge as overcoming the reputation of government work as "unimaginative plodding" while being conducted under a "shadow of military security."[31] Pitzer would have to assure recruited scientists that he understood their concerns and could adequately shield them.

Research was perhaps the most troubling piece of the commission's portfolio, given its essential incompatibility with the military's demands for control and security. Vannevar Bush had largely managed the trick during the war by insisting on OSRD independence from the military, and amid the extreme secrecy of the atomic bomb work, the MED had been able to maintain some level of academic independence due to its close relationship with the university partners managing operations at Los Alamos, Berkeley, and Chicago. The AEC's early leaders attempted to wrest scientific control from the military, with the GAC seeking at its first meeting to maintain "proper balance between freedom of knowledge and positive direction by the Commission."[32] The very structure of the AEC's governance, with the broad input from the academically dominated GAC, guaranteed a strong voice for academic science, and from its earliest years, the commission's main laboratories – Argonne, Los Alamos, Oak Ridge (Clinton), and Brookhaven – partnered with several dozen universities in carrying on their research programs.[33]

Research was critical to the AEC as engineering innovations rested on a continuing stream of basic insights and discoveries. Midway through its first year, the AEC warned that an absence of basic research would cause the United States to yield its "preeminence in atomic energy and weapons."[34] Just as (deliberate) shortsighted planning had created an atomic infrastructure already growing obsolete by 1947, a shortsighted research agenda would produce static knowledge on which few future innovations could sprout. Already the failure to train a new generation of physicists and engineers during the war years was creating a bottleneck in innovation, with virtually all 400,000 of the nation's engineers employed. The GAC predicted a 50 percent decline in graduating engineers between 1947 and 1952, if greater investments were not made in college-level programs, even as demand for engineers was slated to increase. As for physicists, in 1947, the AEC was employing 10 percent of all doctorally trained physicists in the country, and it anticipated hiring more.[35]

Of all of the questions facing the AEC, research in reactor development dominated. The few reactors in existence two years after the war, notably at the University of Chicago's Fermilab (later Argonne) and in Hanford, were all showing signs of obsolescence. Future reactors would need to be smaller, lighter, and safer and above all be able to operate at higher temperatures. Scientists wished to use future reactors to irradiate biological matter, to produce radioisotopes, and to explore neutron capture and optics. Atomic-powered electrical generation would require reactors operating at much higher temperatures (tens of thousands of degrees) which would exploit different construction materials and coolants. Materials would need to be exquisitely pure, lest contaminants compromise reactors' structural integrity as they were exposed to ever higher doses of gamma radiation and neutron flux. And the cooling medium itself – actually the vector for heat transfer in a power-generating reactor – would need to be designed in such a way so as not to corrode the reactor – a common danger in the early reactors.

Since the war, two new reactors had been built: a high-temperature plutonium reactor at Los Alamos and a low-temperature experimental reactor at the new Brookhaven Laboratory on Long Island, NY. By 1949, five new reactors were being planned: one for testing materials within a high-radiation environment, a second designed to fit on a naval ship, a third high-energy breeder reactor, a fourth intermediate-energy breeder reactor, and a fifth being constructed by General Electric in Schenectady, NY for possible power production. The third and fourth reactors, both designed to produce plutonium from ordinary U^{238}, were of particular interest, given the extraordinary inefficiency of the original reactors built during the war. Hanford, for example, lost nearly 99 percent of fissionable material to slag in its inefficient plutonium purification process, and global supplies of uranium were stagnant. More efficient breeding could produce heat (for use in power production) while, theoretically, producing more fissionable material than it consumed. While the breeding process posed certain security threats given its production of (potentially) weapons-grade plutonium, it also marked the only feasible path toward commercial power production in the absence of further uranium discoveries.[36]

The other promising avenue for research lay in newer, larger, and more refined particle accelerators: derivatives of E. O. Lawrence's original 1932 cyclotron, which he had constructed at Berkeley. Such machines were the most promising path to learning more about the forces holding the nucleus of the atom together and about the nature of the subatomic particles in the nucleus. In its early years, the AEC funded some 55 different cyclotrons, betatrons, and synchrotrons, as well as 32 Van de Graf generators, providing broad access for experimental physicists to conduct particle research. In conjunction with ONR, the AEC funded nearly all of the work in this field, which was proving to be the most promising path to understanding some of the basic physical forces of the universe. It also funded the preponderance of doctoral and postdoctoral students working in the field: a total of 660 in the United States in the early 1950s.[37] Nearly half would go on to full-time jobs either within the AEC laboratories or in universities with AEC research contracts.[38]

Infrastructure

As heir to the MED, the AEC possessed or controlled five principal laboratories and three primary production sites. Each of the labs had slightly
different histories. Several were organic enlargements of preexisting university labs (Argonne grew out of the University of Chicago's metallurgy
laboratory, and the Berkeley Radiation Lab had been started within the
physics department by E. O. Lawrence in the 1920s), while Brookhaven was
purpose-developed by the AEC after World War II in an effort to draw on
the research prowess of universities in the northeast.[39] The Clinton laboratory, a spin-off of the Oak Ridge production complex created by the
MED in 1942, had partnered with the University of Chicago during the
war, but afterwards it was orphaned when the Chicago relationship was
replaced by a loose consortium of southern research universities. Ultimately,
the AEC contracted with Monsanto Chemical Corporation, and then later
with Carbide Carbon Chemicals (later Union Carbide) Corporation for
administration, although both corporate partners failed to establish Clinton
as a coequal leg in the Argonne–Brookhaven–Clinton triad. Los Alamos
continued to be administered by the University of California, albeit somewhat under duress.[40] The smaller Knolls lab, in Schenectady, administered
by General Electric, concerned itself with research on power production
while the Ames lab partnered with Iowa State College, mostly on metallurgical research.

The labs tended to take on areas of specialization, although there was
much overlap. Berkeley, for example, became the AEC's lead on cyclotron research, while Brookhaven (along with Berkeley) focused on basic
science, ultimately necessitating construction of its own 184-inch cyclotron. Argonne, built about 20 miles outside of Chicago, was designed to
continue Enrico Fermi's experimental reactor work which had led to the
world's first sustained fission reactor at Stagg Field at the University of Chicago in 1942. Los Alamos dedicated itself (not surprisingly) to weapons
research under the tightest security of any of the labs, while Clinton was
more focused on specific engineering problems. By the 1950s, Argonne and
Brookhaven emerged as the most scientifically formidable, drawing on the
expertise within the academic departments of their partnering universities,
although the Berkeley Radiation lab continued to be the foremost US locus
for elementary particle research.

A more helpful lens through which to view the AEC's research enterprise
in the late 1940s is one of evolving decentralization. While the MED during the war operated under military authority, with centralized command
in Oak Ridge, the new AEC was forced to reconfigure the nation's atomic
effort into a broader, more loosely woven net of uncoordinated research,
exploration, and construction. In an early report of the AEC, chairman
David Lilienthal noted that the commission envisioned a decentralized
organization wherein broad authority devolved to the heads of each of

the five major labs. And while each lab had a general purview, within that purview the lab director could exercise considerable discretion. "The new arrangement," he wrote, "is designed to minimize much of the delay and procedural involvement that might accompany the continued centralization of the vast enterprise either in Oak Ridge or in Washington."[41]

Intrinsic to the AEC's work was its relationship with contractors. Although the AEC owned its facilities outright, it ran almost none of them. General Electric operated Hanford (having inherited it from its wartime administrator E. I. DuPont), Union Carbide ran Oak Ridge, the University of California ran Los Alamos, and the University of Chicago ran the Metallurgical Lab. General Electric also ran Knolls (along with a smaller affiliated facility at West Milton, NY). Each of these primary contractors employed dozens of subcontractors or provided research grants to affiliated universities. Westinghouse, by 1949, was eager to win a portion of the Knolls contract. A new Savannah River plant would be managed by DuPont on a $650 million contract which, in turn, spent $350 million of these funds contracting with over 1,500 other firms. Sandia Corporation was the lead government contractor managing the stockpile of fissionable material. In 1950, the AEC would experiment with a new owner relationship when it granted North Carolina State College permission to operate the first non-AEC-owned nuclear reactor (albeit a small one), although all fissionable material would, of course, originate within the AEC itself.[42] By that time, the AEC employed only 5,000 people directly, even as its contractors and subcontractors employed close to 100,000.

All of these contracting relationships raised the question of private ownership and control of nuclear reactors. From the latter days of World War II, government planners had insisted that atomic energy, and certainly atomic weapons, must remain out of private hands. Whether the mechanism for enforcing that prohibition lay in government monopoly or international control had been decided by the Soviet Union's unequivocal veto of the "Baruch Plan" for the international control of nuclear material in 1946.[43] Pivoting to government control, the United States created the AEC to impose a state monopoly on all aspects of atomic research, fuel production, and weapons development. Clearly, such a monopoly would require some partnership with independent university-based researchers, but government would exercise oversight over these partnerships through funding codicils and mandates for secrecy. That is, the basic condition which the government imposed for atomic researchers was that they work along federally defined lines of inquiry and agree to submit to iron-clad security mandates. While some scientists and universities balked at these mandates, all eventually capitulated; the funding was simply too generous to refuse.

But partnering with corporate administrators seemed born of a different paradigm. Although bound by contractual stipulations with their federal funding, private corporations were owned by their shareholders who demanded that they prioritize profits. Corporations tended to seek new lines of revenue and profit, leveraging technical expertise and specialized knowledge to beat out competition. How was it possible that General Electric, Monsanto, DuPont, Union Carbide, and others could not be exploiting their positions as corporate partners to the AEC to enhance expertise and formulate knew product lines? While they were constrained by US control over fissionable material, the long arc of corporate growth suggested that at some time in the future they would be able to produce weapons or armaments for the government or build power plants for civilian power generation. Both ONR and the AEC could insist on dealing exclusively with US corporations (not difficult to do in an era when both Japanese and European industries were recovering from the war), but federal agencies could not, in the end, manage the disposition of privileged information within these firms, nor their ability to exploit that information to long-term gain. The AEC was redefining notions of ownership and control even as it sought to arrogate ownership exclusively to itself. These patterns of ambiguous control and ownership of intellectual capital would continue throughout the postwar period, adapted for the military, the space program, and the hydrogen bomb.

Notes

1 Quoted in Jessica Wang, *American Science in an Age of Anxiety*, p. 19.
2 Joseph Ball, speech before the Cincinnati Foreign Policy Institute, November 9, 1945, EOL, 108499.
3 An important sideline was the Acheson-Lilienthal report, authored by a committee of five appointed by Under Secretary of State Dean Acheson and chaired by David Lilienthal, the chairman of the Tennessee Valley Authority. The Lilienthal group recommended turning over all nuclear assets and all known uranium reserves to an international body, likely to be placed under the aegis of the United Nations. Bernard Baruch, the US representative to the United National Atomic Energy Commission, proposed the idea to that body in 1946. The Soviet Union, led by its delegate to the United Nations, Andrei Gromyko, immediately vetoed it. See *Report on the International Control of Atomic Energy*, March 16, 1946, widely available online.
4 Wang is quoting journalist Louis Falstein here. See Jessica Wang, *American Science in an Age of Anxiety*, University of North Carolina Press, 1999, pp. 17–18.
5 Ibid.
6 Wang, *American Science in an Age of Anxiety*, p. 18.
7 Richard Niehoff, "Organization and Administration of the United States Atomic Energy Commission," *Public Administration Review*, 8:2, Spring, 1948.
8 The United States Department of Energy ultimately resolved a class action suit from "downwinders" of Hanford for nearly $70 million. See Rebecca Boyle, "Greetings from Isotopia," *Distillations*, October 12, 2017, Science History Institute, https://www.sciencehistory.org/distillations/greetings-from-isotopia.
9 "Control of a Gaseous Diffusion Plant," *Bulletin of the Atomic Scientists of Chicago*, 1:4, February 1, 1946.
10 Lee DuBridge, "Policy and the Scientists," *Foreign Affairs*, April 1963, pp. 576–77.
11 AEC, "Report to the President," April 3, 1947, SRF, 1:1. Over the following decades, substantial new deposits of high-quality uranium ore would be discovered in

many nations, most notably in Kazakhstan, Australia, the United States, and Canada. But the reader should be aware that U^{235} is a finite resource; once depleted it cannot be remade anew.

12 Quoted in "The Atom and the Businessman," *Fortune*, January 1949, p. 57.

13 Karl Compton to Arthur Compton, July 16, 1947, VB, 608: Arthur Compton.

14 Remarks by Commissioner Robert Bacher, March 17, 1949, IIR, 17:4.

15 Wallace to HST, March 15, 1946, PSF, 96: AEC.

16 Vannevar Bush, draft or untitled report, VB, 3091, p. 12.

17 Vannevar Bush, "Statement on Atomic Energy Legislation," December 3, 1945, EOL, 108569, p. 6.

18 Quoted in Wang, *American Science in an Age of Anxiety*, p. 22.

19 Oppenheimer to Truman, December 31, 1947, PSF, AEC Advisory Committee.

20 Ibid., p. 22.

21 From the text of the Atomic Energy Act of 1946.

22 Forrestal to HST, July 21, 1948, SRF, 11A:10.

23 "The Atom and the Businessman," *Fortune*, January 1949, p. 53. The article described the NEPA project as a "Buck Rogers" concept. It ultimately came to naught.

24 Special Committee of the National Security Council, "The Proposed Acceleration of the Atomic Energy Program," undated (1949), SRF, 11A:5, p. 2.

25 Oppenheimer to Truman, June 14, 1952, PSF, 96: AEC.

26 Vannevar Bush, "Statement on Atomic Energy Legislation before the Senate Special Committee on Atomic Energy," December 3, 1945, EOL, 108579, p. 9.

27 Quoted in "Atom Control: Personnel and Policy," p. 3.

28 HST, "Statement by the President," July 24, 1948, SRF, 1:1.

29 AEC, *Annual Report*, 1949, p. 10.

30 GAC to HST, December 31, 1947, PSF, 174: Atomic Energy Advisory Committee, p. 2.

31 "Atomic Boss," *Time*, 53, January 3, 1949, p. 32, as quoted in Robert Seidel, "A Home for Big Science," *HSPBS*, 16:1, 1986, p. 145.

32 GAC, January 3, 1947, as quoted in Ibid., p. 142.

33 Argonne, alone, partnered with 20 universities in 1947, while Clinton partnered with 14 and Brookhaven with nine. AEC, *Semi-Annual Report*, July 24, 1947, SRF, 11A:5, p. 3.

34 AEC Report, July 24, 1947, SRF, 11A:5, p. 12.

35 Untitled document, 1951, DB, FA 965, 28: AEC 1951, p. 9.

36 See Robert F. Bacher, "Remarks at Wellesley College," March 17, 1949, IIR, 17:4. Also, AEC Report, July 24, 1947, SRF, 11A:5.

37 Thomas Johnson, "Remarks at the American Physical Society," April 30, 1953, DB, FA 965, 28: AEC 1952.

38 Statistics are from Boyer to Waterman, April 5, 1951, DB, FA 965, 28: AEC 1951.

39 The two branches of Lawrence's lab, at Berkeley and Livermore, were initially known together as the Berkeley Lab. Ultimately, the two became known as Lawrence Berkeley National Laboratory and Lawrence Livermore National Laboratory; both managed by the University of California. Brookhaven was staffed by a consortium of researchers from the University of Rochester, Columbia, Harvard, Princeton, the University of Pennsylvania, Johns Hopkins, Cornell, MIT, Yale, and Columbia.

40 Initially, the regents of the University of California wanted to terminate their relationship with Los Alamos after the war, for fear of getting overly embedded in classified military activities. Lawrence, himself, played a critical role in drafting the agreement between the AEC and the university to guarantee continued funding for the two labs and ensuring at least some measure of scientific independence. See Seidel, "A Home for Big Science," p. 138.

41 "AEC Report," July 24, 1947, SRF, 1:1, pp. 5–6.
42 "First Non-AEC Nuclear Reactor to be Built at North Carolina State College," October 29, 1950, DB, FA 965, 28: AEC 1950. Notably, the AEC was quick to reassure the public that the new reactor would produce less than one gram of plutonium each year.
43 See note 3.

5 The National Science Foundation

Freedom of Inquiry

The primary recommendation of Vannevar Bush's 1945 report, *Science, The Endless Frontier*, had been that the US government would need to continue its support of the national scientific effort even in the absence of an active war. The extraordinary progress made during the war had been a direct result of government mobilization and direction of the nation's scientific effort, and the major wartime breakthroughs – fission and radar – had required the investment of many hundreds of millions of public dollars.

Bush was most concerned with continued recruitment and development of scientific talent. Alarmed by financial obstacles thwarting "first class" men from pursuing scientific careers, he wrote letters to colleagues around the country emphasizing the importance of attracting men of "keen intellect" who could "do the work." If men of the right temperament were recruited to the field and funded adequately, lack of organizational structure would present no impediment.[1] Other scientific leaders concurred, with Ralph Gerard, a neurophysiologist at the University of Chicago writing,

> The problems for the future are to find and to recognize true talent and leadership, to keep open the road to the top for all individuals possessing these qualities regardless of social status, and to maintain in scientists a greater concern for the good of the many than for selfish gain.[2]

The great barrier to finding and retaining such individuals had always been uneven support. People with the intellectual capacity and drive to conduct scientific research at the highest levels tended to have professional alternatives, and few were so committed to a life in research as to knowingly embrace a lifetime of penury. Industrial labs had long provided an obvious avenue of employment for individuals aspiring to research careers, but applied research careers could not produce the type of fundamental breakthroughs that Bush and his colleagues so valued. Detlev Bronk, at the time a research professor at the University of Pennsylvania and later the president of both the Rockefeller Institute and Johns Hopkins University, explained to a

DOI: 10.4324/9781003363897-5

Congressional committee in Fall 1945, that the biggest deterrent in recruiting the most promising young people into science had been the "unpromising and uncertain economic status of the scientist."[3] Coming out of the war, doctorally trained biologists with a half dozen years of teaching and research experience were finding jobs paying no more than $3,000, with no steady stream of research support funds.[4] This was at a time when senior researchers in applied industrial labs were earning as much as $15,000.

The problem was largely confined to support for basic research, as research with a ready application could generally find a buyer. Industrial labs tended to hire scientists in staff roles and guarantee them standard packages of company benefits such as paid vacations, pensions, and steady employment. Contract researchers, too, led financially stable lives, living off substantial Army and Navy commitments or subcontracts with industrial concerns. In 1946, for example, 90 percent of all federal research and development money was channeled through the War and Navy departments, which tended to renew contracts repeatedly with industry-based researchers. Excluding funding on atomic energy, the military research commitment at the conclusion of the war hovered at just over $1.5 billion while other branches paled in comparison – $11 million from the Public Health Service on biomedical efforts and $37 million from the Department of Agriculture on seed and pesticide research.

Such contracts, however, could not support scientists engaged in independent inquiry on basic questions. Such inquiry required freedom, steady support, independence, and very light administrative oversight. Scientists engaged in basic research developed their own research questions and leveraged their own insights, developed through vigorous engagement with their peers and graduate students, to develop new lines of questioning and novel experimental techniques. Freedom of inquiry was critical to the process. Bronk testified in 1945, "The most important condition for effective research is freedom for the scientist to follow paths suggested by his curiosity or expectedly revealed by his experiments." Funders who demanded that specific problems be solved, or specific applications be produced, risked undermining the essential nature of inquiry. "Scientific discovery is the exploration of the unknown," explained Bronk, "and I, for one, do not see how it is possible to direct an explorer through unknown territory."[5]

The very nature of government contracting, particularly military contracting, seemed to preclude free inquiry given the rigid demands of the federal budget process. Budget requests, as approved either within agency budget offices or within the central Bureau of the Budget, required several lines of approval with stipulated deliverables to qualify for payment. Some agencies had managed to work around federal budget guidelines through extraordinary wartime exigency, but this sort of latitude would be more difficult to grant once the war was ended. Charles Kidd, a staff member on President Truman's Scientific Research Board, explained to a meeting of the American Association for the Advancement of Science, "In federal contracts for fundamental research with universities, as contrasted with development work, the questions of freedom of research and freedom of discussion are always present." And while some of the wartime largesse continued to flow to universities in the months after the Japanese surrender, Kidd questioned the long-term viability of the arrangements. He explained, "One of the problems for the future is to decide at what level the support ought to continue and to establish uniform policies ensuring that the research will be free."[6]

The freedom to pursue scientific work needed to be interpreted broadly. Productive scientists not only thought and worked independently but also published their findings, presented their work at professional gatherings, shared their insights informally with colleagues, and generally engaged with the broader scientific community. Secrecy and loyalty standards were largely incompatible with both the inclination of the scientists and the process of science. Alan Waterman, a former Yale University physicist who was then directing research at the Office of Naval Research (ONR), tried to explain to his military superiors the unique mindset of the working researcher engaged with basic science – the need for "complete freedom of choice in his work, his location, his environment."[7] The scientist was fundamentally an "explorer at heart," explained Waterman. "He wants to pioneer in his chosen field."

Moreover, applied research, with its steadier funding support and more recognizable applications, tended to drive out basic research.[8] The solution must be to fund basic research with a comparable (though not equal) level of sustained support as applied research had commanded during the war. Such support would need a structured basis which allowed some shielding from the highly politicized federal budget process and would need to be shielded from politics generally. Basic science was, by definition, politically vulnerable, with its elitist tinge and absence of demonstrable payoff. It needed to be protected in its essential uselessness, as many societies had historically protected a priestly caste. Bush wrote, "Science has been in the wings. It should be brought to the center of the stage."[9]

Legislative Bickering

The nation's scientific effort had traditionally failed to excite members of Congress. In responding to constituent concerns, Representatives tended to focus on jobs and infrastructure and, to a lesser degree, military preparedness. But the demands of war impelled some members of Congress to consider the role of government in supporting the nation's technological and scientific effort, insofar as more advanced weaponry was coming to play a significant role in military planning. Senator Harley Kilgore (D, WV) was one such legislator, who initially became interested in science and technology while serving on the Senate's War Investigating Committee during the early years of the war.

While serving on the Investing Committee, Kilgore witnessed a procurement error – underperforming synthetic rubber – which could have been avoided had the Army been more technologically savvy. (Engineers had warned of the defects but were ignored.) Later, while serving on a subcommittee of the Military Affairs Committee to study scientific and technological challenges in pursuing the war effort, Kilgore determined that the nation needed a new program to expand and improve its scientific capacity. In 1942, he introduced a science mobilization bill which called for a system of government-sponsored fellowships for graduate study in science and engineering, a new research board which would sponsor federally funded research, and government ownership of all patents emanating from such research. While the bill ultimately failed, it established a framework for postwar government engagement with scientific research. It also framed areas of disagreement between progressive advocates (such as the American Association of Scientific Workers, as well as prominent New Dealers such as Thurman Arnold and Henry Wallace) and scientific elitists – the powerful individuals who led the nation's most prestigious universities and research organizations such as James Conant (Harvard), Karl Compton (MIT), Vannevar Bush (the Carnegie Institution), and Frank Jewett (the National Academy of Sciences).[10] Disagreements between Bush (who had written his report at the direct request of the President) and Kilgore (whose first responsibility was to his constituents) framed lines of disagreement not only between the two branches of government but also between two camps with opposing visions of postwar American science.

At the heart of the disagreement lay the base tension between freedom and accountability. The best basic science demanded a high level of autonomy by the researcher – the freedom to pose a question and pursue a response over many years in conjunction with input from peers, with little or no oversight or accountability. Such a process recognized the essential *creative* component of science. Since the essence of science was finding something that nobody had found before, or doing something that nobody had done before, the process resembled nothing so much as an artist at work, largely following his or her own inclinations and judgment. But science, unlike art, demanded generous funding to pursue, and some of the most exciting branches of the physical sciences now required large amounts of money. Larger synchrotrons, radio telescopes, Van de Graff generators, and research reactors could cost millions of dollars. The funder, usually the government, rightly demanded some level of oversight and supervision. It was simply unrealistic to ask the nation's taxpayers to hand over substantial funds to unaccountable scientists at work in exclusive institutions without being able to hold researchers to task.

Further complicating the situation was that the most innovative science was so esoteric and complex that only other working scientists, themselves highly independent, could reasonably be expected to pass judgment on its potential value. That is, in choosing which projects to fund, the government would be forced to rely on the guidance of independent scientists engaged in active research, who themselves would be reluctant to surrender their independence to federal oversight.

And adding a further layer of complication was the question of disbursal of funds, which could be released to states, regions, institutions, or individuals. Most scientists, of course, preferred that funds be disbursed directly to the most brilliant and creative individual researchers, regardless of physical locale or institutional affiliation. Senators, being subservient to voter concerns, were more inclined to disburse funds broadly to their home states or districts, which could then allocate the funds in some equitable manner to resident institutions and researchers. The first approach promised that the majority of the funds would go to the most qualified and innovative scientists but at the cost of concentrating the funds in the handful of prestigious universities and institutions where the best scientists tended to work. (Bush had previously endorsed this vision of the American scientific organization, which he dubbed "making the peaks higher.")[11] That is, such a system might be good for science but bad for social equity. It would have the effect of making the richest, most formidable, and most prestigious universities that much richer and more formidable and possibly more prestigious.

Kilgore, a populist from the poor state of West Virginia, was more inclined to spread largesse broadly. Moreover, while he and Bush fundamentally agreed on the need for a national science foundation (or a national science board) to fund training and research, Kilgore was more inclined to make the beneficiaries of funds more accountable to the government, including requiring them to surrender any patents which developed out of the funded research to the government and ultimately to the American people. And while Kilgore ultimately opted against channeling money through the states or through third-party national organizations such as the National Academy of Sciences, he did insist on subordinating the authority of the leadership of any new national science foundation to political leaders. Kilgore wrote in December 1942:

> It is for this that the *sine-qua-non* of any Government agency is that its powers be vested in full-time Government employees whose principal responsibility is their public function, and who have severed all previous connections with private financial interests. From the Government standpoint it is unthinkable that the powers of the proposed National Science Foundation be vested in a board of non-compensated persons, whose principal responsibilities would lie in some other direction, as some scientists have so urgently and so honestly recommended.[12]

Kilgore's solution was a full-time board of scientific advisors, serving on a rotating term, working closely with a presidentially appointed executive director who would have ultimate authority over dispersal of funds. Such a mechanism, while requiring the administratively cumbersome device of a scientific advisory board, allowed the government to maintain ultimate control over funds while disbursing them thoughtfully. In balancing the efficiency and accountability of a politically appointed administrator with the scientifically necessary input from working scientists, Kilgore found the compromise solution the most appealing, granting "advisory" capacity rather than binding authority to the scientific board while *requiring* the administrator to meet with the board on a monthly basis and granting direct access by the board to Congress and the President. This arrangement further allowed the administrator to work closely with executive agencies to direct funds to research projects within fields deemed important to the government. Kilgore deemed an alternative solution, in which eminent scientists would be appointed to the board on a full-time basis with greater executive authority, unattractive for the tendency of appointed board members to grow distant from their scientific colleagues.

On the matter of government ownership of patents, which many leading scientists opposed, Kilgore promoted the idea on the grounds of fairness and intellectual freedom. Given that taxpayers were funding the research, Kilgore felt that they should own any useful results of the research much as an industrial concern owned patents from useful devices coming from its own labs. More importantly, giving basic researchers a financial stake in their own work meant biasing them toward lines of research with the most promising immediate applications – the exact opposite of what a proposed national science foundation was designed to thwart. Kilgore and his allies wished to fund scientists working on problems of intrinsic scientific interest rather than on problems with clear applications (which were well handled in either industrial labs or through the military contracting system). Monetary concerns might dissuade scientists from sharing their most valuable findings with their colleagues – the necessary prerequisite for the incremental progress on which modern science was built. "Is there not the further danger that the possibility of patentable results would tend to conflict with free intercourse among scientists in the full publication of research?" wrote Kilgore. "In other words – is not the policy of free dedication the one which most nearly conforms to the ideals and practices of scientists themselves?"[13] Kilgore ultimately embedded his vision for the new foundation in S. 1297, proposed in July 1945 with Truman's endorsement.

Bush disagreed. Concerned that a politically appointed executive director might feel pressure to divert funds to districts of politically powerful members of Congress, he enlisted Warren Magnuson (D., WA) to draft a competing bill (S. 1285) in which power was vested in a part-time board of scientists with no politically appointed director; patent control was retained by scientists and their universities; and funding for social science was notably absent (no surprise given Bush's roots in the physical and engineering sciences). The Magnuson bill found broad support among scientists, a number of whom organized themselves into a Committee Supporting the Bush Report under the leadership of Isaiah Bowman, president of Johns Hopkins

University.[14] Lee DuBridge, the director of the Radiation Laboratory at MIT (and later president of the California Institute of Technology), lobbied vigorously for patent rights for researchers – a stipulation working to the clear advantage of the types of elite engineering schools with which he had been associated.[15]

Elite scientists, or at least scientists working within elite institutions, feared most the potential politicization of science under a politically appointed administrator or at least the usurpation of ideologically pure science. Government could mobilize unparalleled resources (as had been amply demonstrated during the war) but nearly always did so toward political ends. In wooing Isidor Rabi to his camp, Bowman warned of the "capricious control" embedded in the Kilgore bill, "devoid of essential safeguards."[16] While Kilgore viewed the presidentially appointed director as a guarantor of accountability, Bowman viewed such an individual as a scientific weak point, vulnerable to political hijacking. Despite the many signatories to the letter who had benefited handsomely from federal funding during the war (Einstein, Oppenheimer, and Fermi among them), the consensus view among elite scientists was that government could not be entrusted with scientific oversight and that bureaucratic inertia would inevitably run counter to scientific independence.[17]

By December 1945, both Kilgore and Magnuson had withdrawn their bills, and Kilgore had introduced a compromise bill (S. 1720) which strengthened the scientific advisory board but maintained a politically appointed director and held largely firm on social sciences and patents. After some debate, the bill was rewritten yet again and introduced in February of the following year as S. 1850 endorsed by both Kilgore and Magnuson, as well as Senators Leverett Saltonstall (R., MA), William Fulbright (D., WI) and others. At the same time, Wilbur Mills (D., AK) introduced his own bill in the House of Representatives which largely echoed the original Magnuson bill. In the ensuing debate, the delicate compromise coalition which had supported S. 1850 dissolved as Bush withdrew his support and endorsed the Mills bill. Both bills failed that summer.

The failure of S. 1850 left the United States by late 1946 with still no formal structure to support basic scientific research, education, and training. The decline in the ranks of doctorally trained physicists and engineers would continue, with relatively few new candidates entering the seven-year pipeline of PhD and postdoctoral work. And while it was true that the Navy was expanding its network of training grants through its Office of Naval Research, and even developing a grant program in basic research in the physical sciences, the essential orientation of that office continued to be applied engineering work contracted out on a work-for-hire basis.[18]

Not all were unhappy with the demise of the program. The most paranoid of scientists, as represented by Homer Smith, a professor of physiology at New York University, deemed the failed proposal the "largest invasion in history of peacetime government in our intellectual life," – a move which would "wend its way through many complex problems involving private initiative," influence recruitment of the next generation of scientists, and bias science toward the priorities of government.[19] Champions of the 80-year-old National Academy of Sciences (and its associated National Research Council), founded by Congress during the Civil War, questioned the need for any new government funding structure given the time-tested wisdom of the Academy and other respected nonprofit organizations such as the American Red Cross – all capable of ably providing Congress with necessary council in distributing scientific funds.[20] Jewett, of the National Academy of Sciences, saw no reason to establish any sort of national science foundation, suggesting instead that a simple adjustment in the nation's tax laws would produce adequate philanthropic funding of the nation's science.[21] Such reactionaries, however, were very much in the minority. Bush and Kilgore had disagreed on several operational points, but both agreed fundamentally that the nation was in critical need of a substantial and permanent funding mechanism for basic research and training.

Magnuson tried again the following year, submitted S. 526 in conjunction with a new group of cosponsors, notably H. Alexander Smith (R., NJ). The new bill essentially reintroduced Magnuson's original 1945 bill, with a large standing scientific advisory board which would hire an executive director. Patent rights would be retained by the funded universities and researchers, although scientists who actually served on the board would surrender their patent rights to the government. Social science would remain unsupported. Although the bill and its partner bill in the House passed that summer, Truman vetoed it for lack of a politically accountable executive director. In the President's words, it implied "a distinct lack of faith in democratic processes" and judged that it would "impede rather than promote the government's efforts to encourage scientific research."[22] Moreover, the very notion of such powers invested in part-time volunteers, rather than appointed public servants, insulted the President's view of government. "Administration of the law should be vested in full time officers," he wrote in his memorandum of disapproval. Existing bureaucratic structure hardly precluded the government from calling on the services of scientists on a part-time basis.[23]

Scientists, themselves, were fairly divided by this point, with nearly a third who were polled favoring a strong, political director; a third favoring a board-appointed director; and a third favoring a small central executive committee with some executive powers (similar to the AEC).[24] Contrary to Bush's general line of argument, some scientists, particularly those working in the biological and social sciences, feared a science foundation dominated by their colleagues in the physical and engineering sciences who would grab a disproportionate share of authority and the resulting funding. Eaton MacKay, a senior researcher at the Scripps Metabolic Clinic in La Jolla, California, expressed dismay at the Smith and Mills bills (heavily derivative of the Bush plan) for their bypassing political oversight and handing devolving authority to an elite scientific cabal – "bound to be hamstrung by reactionary organizations and sectional controls."[25] Lack of a mandate for geographic distribution of funds meant, in this analysis, that a handful of the most prestigious and highly staffed universities – MIT, Berkeley, Harvard, Johns Hopkins, Caltech, Princeton – would continue to grab the largest share of federal largesse, eclipsing small but high-quality institutions such as MacKay's own Scripps. Already fewer than 50 universities in the United States (out of approximately 1,900 four-year institutions) were receiving over 90 percent of all research funds being distributed by the Army and the Navy, and there was little reason to believe that a new science foundation would do anything other than reinforce these funding inequities.[26] *The New Republic* deemed the whole effort a "triumph for the ivy-league lobby," a reference to efforts by Bush, Conant, and Compton to grab an ever larger share of federal research support for their own institutions.[27]

But, in a larger sense, the multiple failures to create a national science foundation through the immediate postwar years undermined the American scientific effort. Young scientists could not get through graduate school or engage in postdoctoral opportunities while short of funding; innovative ideas could not progress. Leonard Engel, a sharp-eyed science observer, concluded that "every day without a National Science Foundation is a national calamity."[28] The many thousands of young men and women who had failed to receive funding to support their training would likely move on to alternative fields and careers. Engel calculated that as many as 10,000 PhDs and engineers were "lost" to wartime conscription and to the three years of postwar Congressional dallying, not to mention many postdocs and junior researchers.

Moreover, the absence of a national science foundation created a funding vacuum which was rapidly filled by military research. Working scientists needed funding if they were to support themselves and in the absence of a civilian support agency turned their attention to problems of specific interest to the Air Force and Navy. By the end of 1947, the combined services were funding 40 percent of all American research and over 50 percent of university-based research.[29] Research contracts were growing steadily more generous, meaning that whole university departments were formulating their hiring and programs around meeting the research priorities of the military. This did not mean that good work had ceased to transpire, but it did mean that scientists were being steadily diverted from foundational questions to narrower problems with immediate (usually military) applications, often on a classified basis. Engel wrote, "The logical end result of militarized science is the Nazi *Vernichtungslager* [death camp]," with its mechanized processes and synthetic toxins.[30] While Engel's accusations were hyperbolic, his concern over military influence on basic science was hardly misplaced. By late 1947, ONR was funding as much university research as had been conducted in total in the previous year before the war, and even researchers who appreciated naval largesse understood that the priorities of the Navy were not necessarily aligned with the priorities of innovative science.

W. A. Higinbotham, of the Federation of American Scientists, explained that while the policy of ONR had been "highly laudable," more and more scientists were "dismayed at the prospect of increasing military control."[31]

Two years later, in the wake of Truman's reelection, Senators Kilgore and Smith, along with Saltonstall, Guy Cordon (R., OR), Elbert Thomas (D., UT), and Fulbright tried yet again with S. 247. The new bill capitulated to Truman's demands for presidential control over the executive director while the more ardent Bush supporters largely withdrew from debate, sensing that *any* foundation would substantially improve the basic research environment.[32] Truman was so eager to sign an acceptable piece of legislation that he had already submitted a $15 million budget line for a science foundation which did not yet exist. The new bill, with its partner bill in the House of Representatives, passed in late 1949 and was signed into law by Truman in May 1950. Bush and Kilgore's initial vision of a national science foundation doling out both training and research funds to the nation's most promising students and researchers, with substantial guidance from an elite scientific board, had been realized. Funding would be directed almost exclusively to basic research; results would be shared broadly; and patent rights would revert to the government.

Delay had been costly, however. The new NSF would not really become functional for another year: fully six years since the end of World War II. During those years – years of growing alarm over Soviet aggression and atomic advancements – the military (along with the AEC) had become increasingly influential in directing the nation's physical and engineering sciences. In its inaugural year of 1951–1952, the new foundation spent just over $1 million.[33] While that number would grow exponentially over the following decade, the NSF would never really catch up to the defense agencies. Its long gestation and absence had fundamentally diverted American science from a more basic posture, and the nation's premier research institutions had made changes accordingly.

Up and Running

The most contentious issue in creating the National Science Foundation (NSF) was the degree of oversight granted to the President. Having won the ability to appoint his own director, Truman chose conservatively. His pick, Alan Waterman, was coming off a six-year stint as the deputy chief and chief scientist at the Office of Naval Research where he had established precedents and routines of distributing government research funds for scientific grants and contracts. Waterman had earned his PhD in physics from Princeton in 1916, and after a few years with the Army signal corps spent 23 years in the Yale physics department where he rose through the professorial ranks. In 1942, he took a leave of absence to work for Karl Compton at the Office of Field Service of the OSRD. (He had spent a year as a visiting professor at MIT in 1937, where he got to know both Compton and

Vannevar Bush.) He never returned to Yale, going directly from the OSRD to his position with ONR, where he was instrumental in enlarging and strengthening the agency. By the time Truman tapped him for the NSF job, ONR was sponsoring some 1,600 projects per year through a variety of granting and contracting mechanisms.[34]

Waterman was the ideal pick as founding director. Well-respected in both academic and government circles, and with nearly a decade of close association with the military through his work at both OSRD and ONR, Waterman inspired confidence and trust from the three primary constituencies with which the NSF would work: civilian scientists, the executive branch, and the military. In particular, Waterman was greatly attuned to two challenges which loomed for US research: the need for greater investment in training the next generation of scientists and the need to balance the requisite independence of the basic researcher with the applied needs of the military (and the broader economy). At ONR, Waterman had deployed the substantial resources of the US Navy to recruit talent, fund training, provide grants to basic researchers, and align research with Navy goals. Upon Waterman's appointment, Isidor Rabi, the chairman of the ODM and Nobel laureate in physics, offered a ringing endorsement: "The general health of our research in the universities and elsewhere, I think is largely due to the efforts of the ONR, and they cannot be given too much credit for that accomplishment."[35]

The NSF, however, had been envisioned as a *foundation* rather than an *agency*, despite its lack of an endowment, its dependence on Congressional funding, and its political tethering to the White House. As a foundation, it worked through a collegial governance process in which funds were disbursed through the collective judgment of its National Science Board – the 24-member board which actually set priorities and made funding decisions. In practice, the director and his staff would propose fundable projects while ultimate judgment on those projects was reserved to the board. In creating this "peculiar, two-headed structure" (in the words of science observer Dael Wolfle), Congress had to replicate the very structure of the US government (with its divided legislature and presidency) as well as other independent agencies such as the Federal Reserve and the Federal Communications Commission.[36] The structure was designed not for efficiency but for inclusion and debate. Since there was rarely an optimal project to fund, or a correct set of scientific priorities, the success of the foundation would necessarily rely on eliciting the voices of many qualified individuals to articulate different insights and priorities.

In fact, part of the motivation of creating the National Science Board was to bring both scientific *and* political concerns to the fore. Truman's initial appointments to the board included some of the most eminent individuals in American research but also included representatives of different fields of science, different parts of the country, and even different ethnic and racial groups. The first board included leaders of the country's leading research universities and institutes (Johns Hopkins, Harvard, Rockefeller, Carnegie, Caltech, Princeton, Washington University) but also representatives of large state schools (Wisconsin, Wyoming, Missouri, Purdue), Catholic colleges (Spring Hill), historically Black colleges (West Virginia State College), and industry (General Electric, Homestake Mining, and others).[37]

The divided structure would prove to be cumbersome, and over time authority shifted to the director and his large staff, with the board often doing little more than rubber stamping funding choices.[38] Offsetting this trend toward hierarchical administration was a peer-review structure which Waterman implemented almost from the beginning, in which he asked university-based scientists to critique various grant requests; the written critiques then became part of the full application presented to the board. While the reviewers' critiques were not binding, their authority gained stature over time, such that within a few years of the foundation's founding it became highly unusual for either the director or the board to overrule strong negative outside reviews. In effect, the role of the peer reviewers displaced the envisioned role of the National Science Board, which became more removed and passive – more of an advisory body than an active decision-making board. By 1962, Congress formally recognized this evolving structure by converting the board to a general advisory committee and creating a new Office of Science and Technology to take on at least some of the functions of the now moribund board.[39]

From the onset, the NSF was clearly driven by *basic* research, in contrast to the emphasis on applied research embraced by both ONR and the AEC.[40] The Steelman Report of 1947 had stipulated this and in fact had recommended that any basic research funded by ONR be transferred to a future NSF upon its creation.[41] Since the preponderance of the nation's basic research was conducted in its universities and research institutes, the foundation's mandate could be accomplished only through distributing funds either to universities which hosted research programs or to the researchers themselves. While university presidents advocated for channeling funds through the institutions (to be distributed internally), Waterman was inclined from the start toward direct funding of *projects* (as he had done in his grant-making at ONR) which, in turn, meant directing funds toward the individuals who would propose and direct the projects. That is, NSF funds would be granted to specific individuals, albeit almost invariably those with university-standing. The policy hardly barred the strongest universities from winning a disproportionate share of NSF grants, but it did raise questions surrounding ownership of data and patents. Research universities began to address this question in 1952, when representatives from MIT, Caltech, and the University of California met with Waterman to resolve differences between institutional policy and foundation policy. In effect, when fellows or senior scientists brought funds with them to a university, it raised the question of which set of rules would now govern those funds – NSF rules or university rules?[42]

A second question facing Waterman was the weight that military (and other government) priorities should bear on funding decisions. While the NSF was envisioned as a basic research funder, that basic research needed to be done with an eye toward the public's long-term interest. Waterman's very appointment had been a nod to the real concerns of the DoD that in matters of national security the foundation's research agenda must capitulate to national exigency.[43] Waterman emphasized this role in his first statement of national research priorities in 1951 which placed "basic research for defense" foremost and which included research projects associated with missile guidance, supersonic and high-altitude flight, and developing compounds which could withstand the high temperatures of jet engines (such as titanium). Such a list hardly seemed an endorsement of the basic research orientation of the new NSF.[44]

Military need was not the only tug on NSF largesse. From the first year, Waterman was aware of other lines of basic inquiry which might lead to practical applications, even if not explicitly contracted to do so. New developments in capturing and utilizing solar energy lay foremost on his list of nondefense priorities, along with harvesting algae (for food) and synthesizing plant-based proteins.[45] Waterman would reassure his congressional critics that the NSF was acting responsibly as a steward of public funds, investing in research which showed promise of improving the security of the nation.[46]

The third question facing Waterman was support for training, which he viewed as intrinsic to the foundation's goals. Grants supporting both predoctoral and postdoctoral work were included in the NSF budget from the first year; within five years these had expanded to support for high school science teachers and for international science education consortia.[47] Unexpectedly, it was the postdoctoral grants which would prove to be the most critical line of support for emerging researchers rather than the undergraduate and doctoral support oft spoken of in Congress. Most universities were able to fund adequate undergraduate science training through tuition revenue, and graduate (doctoral) students could be funded through teaching obligations. But the odd gap between completing the PhD and actually becoming a productive scientist required tailored interim support in the form of the emerging postdoctoral fellowship. NSF support for these fellowships grew rapidly during the early years of the foundation and ultimately became the prevailing model for developing scientific talent.[48]

One of the troubling side effects of increasing federal support for basic research was the manner in which the funding exacerbated inequities among the nation's universities. In 1950, shortly before the NSF's beginning, just 11 American universities accounted for half of all federally funded research (predominantly through ONR and the AEC but also through the nascent NIH) while 50 universities attracted over 90 percent of all federal research funding.[49] While these universities were distributed throughout the United States, they were disproportionately located in the Northeast and on the Pacific coast, while the South was consistently underrepresented. This distribution of funds threatened to weaken respectable mid-level colleges and universities, whose most talented faculty (and possibly students) could be skimmed off by the newly enriched research powerhouses. Waterman recognized the danger but ultimately could not bring himself to withhold support from scientists at the traditional research centers, committed as he was to funding individuals rather than institutions. He admitted in an interview in 1951 that supporting individuals and groups who were best able to complete a particular research project would produce the greatest scientific progress but would tend to narrow the nation's research base. But beyond seeking unrealized "potential research energy" rather than the already-in-motion "kinetic research energy," he offered no solution.[50]

The NSF's emphasis on supporting projects rather than institutions contrasted with research policy in most other countries. While research support in western Europe in the late 1940s was relatively scant, what support there was flowed to regions or universities. France, the United Kingdom, and West Germany all ran research funds through university-based research committees, which allowed the government to either distribute largesse broadly (France and the United Kingdom) or deliberately strengthen a national research institution (such as the Max Planck Institute in West Germany). While the American approach privileged the scientist over the institution, it ultimately made the best American universities even better by further concentrating the most promising researchers at a small group of coastal and mid-west research universities. Bush had predicted this in his 1944 report, predicting, "no matter on what conditions money is given to universities, the very existence of such support will, of course, modify university policy."[51]

Universities which were already strong might have pursued research-active faculty for the sheer intellectual heft they added to faculty rosters, but NSF policy on fringe benefits and indirect expenses added impetus. In 1947, President Truman had sanctioned an ONR policy giving universities the ability to set their own salaries (embedded in ONR contracts and grants) and to make good-faith estimates of indirect expenses associated with grant- and contract-funded research. Moreover, the government recognized a model of "indivisibility" in which teaching expenses were deemed inseparable from research expenses. That is, the professor was treated as a single productive unit whose teaching and research activities were intermeshed. What this meant, in effect, was that a professor applying for an ONR (or AEC) grant could put in place the complete cost which he or she posed to the university, including salary, fringe benefits, pension obligations, office and lab space, administrative overhead, and fair-market applications of infrastructure such as lab buildings, the library, and teaching lab space. Indirect expenses could quickly climb to 70 percent of the cost of a grant, particularly in those fields of research which made use of very expensive pieces of equipment (such as reactors and cyclotrons). In a certain sense, ONR was funding the institution but through indirect additions to individual researchers' grants.[52]

At the outset, the NSF largely adopted existing guidelines for reimbursing indirect expenses, albeit with some caveats surrounding funding underlying faculty lines.[53] The NSF was willing to make long-term (5-year) commitments to complex research projects involving substantial capital investment, though it was unwilling to fund a research line at a particular institution into perpetuity. When equipment became too expensive for the NSF to fund at a particular university, it agreed to fund consortium-based projects in which several universities banded together to do work with an accelerator, a wind tunnel, or a mainframe computer. During its first decade the NSF would fund construction of a large radio telescope in Green Bank, West Virginia, and an astronomical observatory at Kitt Peak, Arizona – both joint programs with multiple universities. It made similar commitments to biological field stations in California (White Mountain), Missouri (Prairie), and Massachusetts (Woods Hole).[54]

Despite the inspired pick of Waterman as its founding executive director, and the solid systems for evaluating and awarding grants, the NSF had substantial hurdles to overcome. The military orientation of wartime research under the OSRD, followed by the explosive growth of ONR (and later the AEC) in the immediate postwar years, left the nascent NSF in a weak position. By 1951, the nation was spending $2.5 billion annually on research, of which 60–70 percent was funded by the federal government ($1.7 billion). Nearly two-thirds of all research was conducted in industrial laboratories, leaving universities with a scant 10 percent, of which only a small portion was basic research. That is, the NSF was established to grow what was a very small part of the nation's entire research effort.[55]

The foundation's budget for its first year was a scant $1.1 million: a pittance compared to the research budgets of the DoD ($178 million) and the AEC ($121 million). Even within the much smaller world of basic research, NSF grants constituted just 1.5 percent of all federal support for basic research that year. NSF staff members who spoke with university-based scientists throughout the country during the early years of the foundation found that many researchers were unaware of its existence or confused it with the National Academy of Sciences.[56] Congress, in particular, was unenthusiastic about the new foundation; in 1952, the House Appropriations Committee reduced the foundation's budget request from $14 million to $300,000, refusing to fund both training and research grant programs in light of the escalating Korean War. Waterman urged the committee to rethink its funding decision, and to recognize the critical long-term role in national preparedness that the foundation would play, but the Senate committee held fast.[57] In 1953, the NSF's research and training budget rose to only $2 million, compared to $1.7 *billion* spent on research by the DoD.[58]

Waterman was appalled by Congress's myopia, pointing out that extraordinary wartime success had rested on earlier basic research and even earlier investment in education and training. Speaking to the Senate Appropriations Committee, he urged members to recognize that research was "difficult,

unpredictable, and hazardous," and that it was predicated on training competent scientists in a milieu of basic research – the exact effort which Congress
was so aggressively undermining. "It must be a source of concern to all of
us," Waterman continued,

> that while the number of students graduating in the sciences in the
> United States continues to be inadequate, in the Soviet Union it appears
> to be rapidly rising. Greater support for basic research will help increase
> our output of scientists at the same time that it adds to our store of sci-
> entific knowledge.[59]

While the foundation's budget would slowly rise through the 1950s, its
failure to gain initial traction delayed its impact on the nation's basic research
and biased scientific efforts toward military applications. The gigantic DoD
Research and Development Board (RDB) dominated all grant-making
activity in Washington (with the partial exception of the AEC and the
NIH, both of which had more narrowly defined purviews). Waterman was
fighting sheer inertia in Congress, which well understood the role of the
military but had not yet grasped the potential contribution of basic research
to future progress. But the NSF was hurt, also, by reports of disloyalty in
university ranks (a disproportionate share of avowed American communists
could be found in university faculty by the mid-1950s) and by bureaucratic
sabotage by the Army and the Navy, which failed to understand why the
NSF's function could not be easily folded into the RDB. At least one science newsletter publicly questioned whether or not the nation might not be
better served by a cabinet-level Department of Science, although neither the
White House nor Congress seriously entertained the idea.[60]

Effects on Universities

One of the initial challenges confronting the NSF was that of distribution.
In 1951, the nation had approximately 1,900 four-year colleges (with about
half that many again of two-year or "junior" colleges). These colleges, in
aggregate, conducted about 90 percent of the nation's basic research and
trained virtually all of the nation's engineers and sciences at both the undergraduate and graduate level, with many hosting postdoctoral fellows as well.
But within the group lay enormous differences, with faculty at many of the
colleges performing virtually no research, while faculty at a few dedicated as
much as 80 percent of their time to research. Universities and colleges (colleges had historically hosted no programs beyond the bachelor's level, but
more were adding select master's programs during this time) often explicitly typed themselves as one of several types of loosely defined institutions:
liberal arts colleges, professional universities, research universities, technical schools, and agricultural and mechanical institutes (often holders of the
state's land grant). Research universities, often with their accompanying

medical schools, performed the most research, although agricultural and mechanical institutes performed a fair amount of agricultural and engineering research, and technical institutes tended toward applied research.

But even within types lay huge differences. A handful of research universities performed substantially more research than others. ONR determined in the late 1940s that probably no more than 700 universities in the country had the infrastructure, resources, and faculty talent to conduct research worthy of federal funding, and that in the year before the NSF was created only 225 of these 700 received any federal research funds at all. And even among these 225, there were enormous differences in research capacity, with just five institutions receiving 55 percent of all federal research funds.[61] Research in 1951 was an elite sport, dominated by a few giant players which collectively had such a head start on their competition that it was difficult to see how the government could have an equalizing influence.

The problem with such institutional inequity was that the bipartite mission of the NSF, research and training, suggested different strategies in distributing funds. High-quality research tended to aggregate itself; the smartest, most innovative researchers wanted to be around their intellectual peers where they could work together on large projects with tricky solutions. Given that many universities hosted virtually no research, investing scarce federal research funds in these institutions seemed counterproductive; there was little chance that they would be able to remake themselves into productive research factories or be able to sustain hiring practices to attract high-level talent.[62] Harvey Brooks, a physicist who spent a number of years advising Presidents Eisenhower and Kennedy on scientific issues, pointed out that research shops needed to be of a critical size to attract and hold talent which could then collectively win grants to keep the shop vital. "Too much dispersion leads to trivial and unimportant research," he wrote.[63] And, in fact, critical size was actually increasing during these early years of the NSF as projects in the physical and engineering sciences required progressively more expensive equipment and larger teams to run experiments. Not surprisingly, grants made during the early years of the NSF tended to go to large institutions with established research records. The NSF, following a well-worn path, was tending to make the strongest universities stronger still.

But the second part of the NSF's mission, teaching and training, tended to pull grant-making toward *broader* distribution. If the United States were to catch the Soviet Union in developing the next generation of talent, it would need to vastly widen its pipeline of scientifically oriented students. In 1950, the nation had just under 800,000 trained scientists and engineers – about one half of 1 percent of its population. The Soviet Union had more, and its universities were producing about 40 percent more engineers each year than were American ones. One science writer warned, "The fast approaching bottleneck of too few scientists and technologists can well be the most efficient weapon possessed by Stalin and the Politburo."[64] The cohort of scientists who had come of age during the Great Depression had been unusually small, meaning that the United States would need to find ways to identify more nascent talent, prevent high school attrition, maintain promising students in college, and support these students through graduate school. Children without college-educated parents disproportionately failed to attend college themselves, and these patterns tended to be laid down early in a child's education. Vannevar Bush noted,

> There are talented individuals in every segment of the population, but with few exceptions those without the means of buying higher education go without it. Here is a tremendous waste of the greatest resource of a nation – the intelligence of its citizens.[65]

While the United States had raised the literacy rate in the general population to near 100 percent by the end of World War II, it faced very high attrition rates through the junior high school and high school years. Only 40 percent of American boys graduated high school in the late 1940s, and only 7 percent earned a bachelor's degree. Fewer than a quarter of American high school students took a class in basic algebra, fewer than 5 percent took a physics class, and fewer than half of all American high schools offered even one course in chemistry or physics.[66] By contrast, in Soviet high schools in the late 1940s, 40 percent of all curricular hours were devoted to math and science.[67] A comprehensive effort to increase the number of American scientists and engineers would need to start with better and more broadly distributed high school science teachers who could move promising students down a scientific path. One NSF staff member, emphasizing the need for broader distribution of teaching and training grants during the 1950s, explained, "It is clear that in order for a teacher to transmit knowledge in science and mathematics, he must as a minimum know something about science and mathematics."[68]

But the calculus was even more complicated. Although research was largely conducted at a handful of large doctoral universities, many of the nation's finest students attended small liberal arts colleges which made few research demands of their faculty. Graduates of these small colleges went on to pursue graduate work in substantially higher rates than graduates of the undergraduate programs at large research universities. In 1951, of the 50 schools in the United States producing the highest percentage of graduates pursuing doctoral work in the natural sciences, 39 of them were small liberal arts colleges.[69] Thus, the work of the NSF was really pulled along three axes: research funds for faculty at the most research-intensive universities who could best marshal faculty and resources to successfully complete large projects; teaching and training funds to be broadly distributed among programs likely to train future high school and junior college science teachers who would be critical in enlarging the pipeline of future students of science; and smaller research grants for teachers at excellent liberal arts colleges (or regional master's universities) where many of the next generation of doctoral students were receiving their undergraduate educations and thus needed to be introduced to the techniques and approaches of a basic research lab.

The essential tension between concentrating research funds in elite research universities and distributing teaching and training fellowships to lesser schools troubled NSF staff members and board members during the initial years. Patrick Yancey, a professor of biology at Spring Hill College, a small Catholic college in Mobile, Alabama, who served as a member of the first National Science Board, expressed concern that the foundation's disproportionate targeting of established research schools undermined a principal goal of the agency: "This was not what the Congress had in mind setting up the Foundation," he wrote.[70] High dropout rates and lost scientific talent could be ameliorated only by creating opportunities at substantially more colleges and universities around the country which could train the next generation of science teachers or could lure promising collegians to graduate school by giving them an early experience in lab research. "The problem of dispersion versus concentration is a real one," wrote one White House science advisor.[71] Waterman concurred but initially felt compelled to target grants at researchers and institutions whose records reflected high rates of research success. And although by 1960 he would claim to scatter support on a "wide geographic basis," and with "special attention to promising young investigators," grant-making research priorities would continue to impel a narrower, more elite focus.[72]

However, despite the compelling need for training science teachers and impelling college students into science careers, the central mission of the NSF as seen by most members of the National Science Board remained basic research support. This was the task on which Bush had focused his 1945 recommendations, and this was the national need which had propelled the NSF legislation through four failed bills, countless amendments, and five years of congressional debate. Bush's vision, leveraged through his national reputation and broad professional scientific network, was the guidepost for the nascent NSF. As early as October 1945, he had warned Congress that although applied science had produced such brilliant results during the war, it could not sustain American leadership. Of his wartime successes at the OSRD, he preached,

> This is the way to win a scientific war, but this is not the way to advance the frontiers of knowledge. The inquiry into the unknown is a vastly different proposition. The research spirit, an intangible most difficult to cultivate and preserve, must be encouraged in many places where it now hardly exists.[73]

But how to discern worthy scientific projects? The very nature of basic science suggested innovation, lack of obvious impact, and absence of clear application. Grant reviewers could examine a scientist's academic qualifications or standing within the scientific community without being able to discern the merit of his or her proposed project. Harvey Brooks, one of Eisenhower's science advisors, pointed out that talented scientists might become excited about an idea that was little more than an "intellectual puzzle having little relation to the real world." He asked, for example, "How really significant is the continued elaboration of nuclear energy level schemes?"[74] While such research ultimately led to profound discoveries into the basic structure of matter, at the time it appeared to be an intellectual dead end.

Brooks helps us to understand the quandary of intelligently funding basic science, as he had participated in science as a scientist, national lab administrator (he served as the deputy director of the Knolls Atomic Laboratory in the late 1940s), academic administrator (he served as dean of engineering and applied science at Harvard for almost 20 years), and political advisor (serving on the Presidents Science Advisory Committee under Presidents Eisenhower, Kennedy, and Johnson). His many roles forced him to confront the troubling challenges of funding basic research, knowing that the future impact of basic research could rarely be discerned at the time that it was undertaken. Part of the challenge could be solved by simply funding good people – those individuals recognized by their peer scientists for their superior intelligence and insight and previous research successes – but the danger here was the best people might be drawn to a particular field of study simply because that was where funding support had drawn them. Inorganic chemistry, for example, appeared moribund in the 1950s. Was this because of a lack of compelling research questions or a lack of innovative and creative scientists who had failed to be recruited to the discipline? It was difficult to know. Moreover, Brooks worried that discoveries could be over-celebrated ("Are we making a fetish of originality? Do we value trivia above an important synthesis?") or underappreciated ("What seems trivial to the outsider may in fact be very important").[75] The task was daunting.

Through the 1950s, Brooks produced a framework on which to judge basic science, calling on evaluators to grade proposals on "generality" [generalizability], applicability, originality, and educational value. Generality meant relevance – the ways in which the research spoke to central questions of the discipline. "To what extent do the results lead to new general principles of wide application?" demanded Brooks. Applicability spoke to basic research as a precursor of applied research: Could the basic investigation ultimately lead to real-world returns, and what was the time-horizon that a funding panel could demand for these returns? Originality was perhaps the quirkiest: both deeply important but potentially tangential to the ultimate course of science. "To what extent is an idea worth supporting just because it is different?" Brooks asked. "How much support should be allocated to projects which pre-dominant expert opinion believes will be unproductive?" A list of exciting new developments in 1960, such as superconducting magnets and viral origins of cancer, had begun with investigations which defied prevailing scientific paradigms. As for educational value, Brooks asked evaluators to consider the pedagogical value of participating in innovative work simply to master modes of thinking or to lure future scientists to the discipline. The connection was misleadingly clear and often falsely claimed. After all, many of the best young graduate students had received under-graduate educations without exposure to innovative lab work.[76]

Through its early years, the NSF attracted more funding from Congress (after its disastrous initial two years), hitting $40 million in 1957, and $136 million in 1959, which allowed it to fund just over a quarter of experimental funding requests of proposals deemed "worthwhile" by 1958 – roughly comparable to the funding rates of other federal funding agencies.[77] Increased funds allowed it to build its fellowship program through this time, for which it created four categories: predoctoral (working toward the PhD), postdoctoral (having recently completed the PhD), senior postdoctoral (greater than five-years out from the PhD), and science faculty fellow (mid-career instructors and lecturers in the basic sciences). These four programs were each designed to provide a livable salary for the fellow (rather than to support a particular research project) in hope that removing financial concerns or teaching obligations would allow fellows to devote themselves to research or to revitalizing their teaching. In effect, the fellowships provided the fellows with time. In addition, the NSF created a series of summer and academic-year institutes open to high school science and math teachers, aimed at presenting new developments in science and new pedagogical techniques which could be incorporated into a high school science curriculum.

Expenditures on fellowship and institutes started at very low levels but jumped sharply in 1956 after Congress increased the foundation's budget allocation.[78] In 1957, the NSF spent nearly $11 million on the programs, which jumped to $13.5 million the following year. Graduate school enrollment started increasing in 1956, possibly in response to fellowship support; the number of enrolled doctoral students doubled over the following decade and tripled by 1970. Moreover, the fellowship programs were proving to be highly competitive, with only 12–16 percent of all applicants at the predoctoral and postdoctoral levels winning funding. Although it was difficult to objectively grade the quality of the applicants, the NSF reported that successful applicants in 1956 had a median graduate ETS quantitative score (the test which would eventually become the Graduate Record Exam) of 740, compared to a national median for all successful graduate students of 490; scores for students in the non-life sciences were higher still.[79]

But here, too, issues of distribution plagued the foundation. In its first class of predoctoral fellows, nearly ten percent chose to attend MIT; hardly a propitious start to expanding access to science education nationwide.[80] The stated purpose of the fellowship program, supporting graduate education for the 10,000–15,000 capable students eschewing graduate work annually for reasons of fiscal want, would require scaling the model to make high-quality graduate education available throughout the country.[81] MIT and its prestigious peer institutions could not accomplish the task alone.

But perhaps the NSF's greatest hurdle in its early years was competition posed by other government funding efforts. The DoD and AEC's massive contracting enterprises had become such important sources of funding for premier American research universities during the late 1940s that the programs had actually distorted the shape of university life. Whole departments had been created and labs built based on continued government contracts. Professors were being hired with no expected teaching obligations, whose salaries could never be underwritten by tuition revenue. Some universities – notably Chicago, Johns Hopkins, Princeton, and Berkeley – were building whole new campuses to house these contract research programs, populated by scientists with only scant ties to the host institution, no undergraduates, and few graduate students.[82] The enlarged contract research programs threatened to draw university attention from core educational activities and to divorce the work of young scientists from basic inquiry. Waterman wrote, with concern, that the lure of contract funds undermined basic fairness across the faculty, accelerating research output for scientists working on military problems while diminishing the work of basic scientists, as well as faculty working in the humanities and social sciences:

> What to do about the situation has become a major question. Obviously the national security comes first, and all of the research and development that can effectively contribute to that end should be done. Nevertheless it may be pertinent to ask whether, in fact, the huge sums presently being appropriated directly for military research and development have taken adequate account of the need for advancing the frontiers of science and for training young scientists and engineers. . . . Both the universities and the federal government should critically examine the question whether all the research that the universities are being asked to do is properly the kind of research that belongs in the universities, or whether some of it might not more properly be carried on in Service laboratories or by industrial organizations.[83]

Defense Spending on Basic Research

Paradoxically, the nascent NSF found itself challenged by the substantial federal commitment to basic research *already in place* at the time of the foundation's establishment in 1951. As has already been noted, the federal commitment to research of all types in the early 1950s, built on the substantial wartime base, was quite massive – about $2.1 billion by 1953. Just over $1.9 billion of that commitment came from the combined research budgets of the DoD and the AEC, with the remainder contributed by the National Advisory Committee for Aeronautics (NACA), the Department of Agriculture, and the Federal Security Agency (which included the Public Health Service). And while the bulk of that research commitment went for contracted applied projects, even the small amount which went for basic research – about \$140 million, or just over 3 percent of the total – dwarfed the budget of the NSF research program during its early years.[84] The emphasis, of course, was on defense needs.[85]

As discussed in the previous chapter, some of this effort came through the AEC, but the preponderance of it was organized and overseen by the RDB of the DoD, the agency most clearly designed to continue the work of the wartime OSRD in the Cold War era. Originally created through action of the Secretaries of the Army and Navy as the Joint RDB, the board was later folded into the unified DoD in 1947; the "joint" was dropped from the title and the chair of the body was given direct reporting access to the Secretary of Defense. Ultimately, the RDB structure was dropped entirely for a dedicated research division within DoD with its own Assistant Secretary of Defense.

The military, by design and inclination, was hierarchical, and its research effort reflected this. Research contracts were structured on a work-for-hire model with clear expectations and defined deliverables. The need for high relevance and applicability drove the military to retain ownership and control over the bulk of the federal research effort for the next half-century. The military recognized, however, that general staff officers and civilian political appointees were unlikely to be familiar enough with scientific and technical problems as to be able to constructively dole out DoD research funds and thus drew on an extensive network of advisory bodies and commissions (staffed with technical consultants) which created study groups and advisory panels to evaluate specific military needs and define the research programs which might be most productive. The process, while ultimately responding to the military chain of command, was designed in such a way as to engage knowledgeable experts, inviting debate and brooking dissent.[86]

The scale of defense research by 1952 was massive, requiring an estimated 54,000 research scientists and engineers or nearly half of the scientists and engineers in the country, with the research needs of the AEC and NACA drawing an additional 20 percent.[87] Growing US involvement in the Korean War impelled greater defense commitment to research, as the military became more aware of innovative Soviet military technology. Secretary of Defense Charles Wilson exhorted the public and Congress to maintain its commitment to new weapons systems, inveighing:

> We must keep our weapon systems ahead of all potential enemies in time of peace with the hope that through being prepared we will deter a possible war. The interval between the onset of aggression and destruction is so short that we cannot afford to delay our development of new weapons until an enemy strikes. We must recognize that the weapons required to repel aggression cannot be and are not conceived and developed overnight. It takes months and years to perfect and produce the new weapons and counter-measures which we are depending on to insure our safety.[88]

At the time, the DoD was funding nearly 8,000 separate research projects through either grants or contracts, with additional research-like projects and funds buried in other divisions of the agency such as the Air Force R&D Command. Notably, in 1953, the Air Force, the Navy, and the Army each was running independent rocket development programs uncoordinated by the RDB.[89]

Even amid this considerable defense commitment, defense escalation of *basic* research grew much faster. From 1945 to 1953, while federal research contracts with universities almost doubled, federal commitment to university-based *grant*-funded research grew 14-fold (from $10 million to $140 million), absent an NSF until the final year.[90] The growth was entirely intentional, with senior staffers within the RDB and throughout the general staff rejecting the notion that the NSF ought to supersede the DoD's role in funding basic research of interest to the military. The military simply did not trust a civilian-led agency informed by a board of civilian scientists to fund the projects of greatest relevance to defense needs. The Science Advisory Committee of the Office of Defense Mobilization (ODM) within the White House suggested that trusting the NSF to adequately fund projects of interest to the military was a "dangerous fallacy" which would lead to underfunding the basic research underlying weapons development.[91] Air defense radar, missile guidance systems, and submarine detection technology were all built on basic research, which had been funded by ONR grants through the 1940s and 1950s. For the DoD, its grant-funded and contract-funded research needed to be carefully coordinated and comanaged. Secretary Wilson made clear his conviction in 1953, a year after the NSF's founding: "We have taken no steps to de-emphasize basic research."[92]

DoD policy in this regard was not wholly at odds with the scientific mainstream. Scientists who had worked on Army and Navy projects both during and after the war had welcomed the generous funding but had also sometimes come to appreciate the importance of coordinating even cutting-edge science with practical need. The ODM's Science Advisory Committee was peppered with world-class researchers such as Oppenheimer and Rabi. Rabi, in particular, appreciated the constructive nature of a close working relationship between client and researcher, specifically between the military services and civilian scientists. He wrote,

> Like our most advanced technological industries . . . the military services must have a class coupling between basic research and technology. Their problems are so large and so complex that it would be military folly to depend wholly upon other agencies.[93]

The military's continued substantial commitment to basic research raised the question of the precise role of the new NSF. The initial $15 million budget cap on the NSF paled next to the combined basic research budget of the DoD and the AEC which rose to $150 million by the mid-1950s. While Bush's original vision for the NSF, penned in 1945, suggested that the foundation should ultimately displace the military in basic research, this was not to be.[94] A 1954 executive order suggested that the NSF would be increasingly responsible for "general-purpose" research while the other agencies (primarily the DoD, the AEC, and NACA) would take the lead in funding "mission-based" basic research, but the order failed to address the fact that fully 95 percent of all basic research at the time continued to be mission-based. Rabi, in fact, urged the DoD to not define its mission too narrowly, lest it fail to fund more general-purpose research which could possibly be of tangential interest to the agency.[95]

The evolving NSF continued to be a much junior partner to the defense agencies through the 1950s and beyond. Emerging from a tempestuous multiyear legislative battle, the new agency was too small and too marginal to the nation's Cold War security concerns. Nearly three-quarters of the nation's research scientists and engineers had worked on DoD or AEC-funded projects by the late 1950s, and most had found the relationship conducive to good science. Along with senior White House and DoD officials, they were reluctant to undermine a large system which appeared to be working well and was respected by American voters. Historian Bruce Smith writes, "Support of basic research by the agencies was a technical overhead on their missions, and the NSF budget was an overhead on the overhead."[96] The NSF, initially, was less a threat to the defense research programs than a curious sideshow; a bit idealistic but largely irrelevant to the pressing work at hand.

Notes

1 VB to Karl Compton, September 13, 1946, VB, 609/Karl Compton.
2 R.W. Gerard, "A National Science Foundation and the Scientific Worker," *Science*, 103:2662, January 4, 1946, p. 5.
3 Detlev Bronk, "Testimony to the Senate Committee on Commerce and the Senate Committee on Military Affairs," October 24, 1945, DB, FA 965, 55: DWB speeches.
4 Ibid.
5 Ibid., p. 8.
6 Charles Kidd, "The Federal Government and Shortages of Scientific Personnel," December 28, 1945, SRB, 6: Scientific Personnel, p. 3.
7 Alan Waterman, "Fundamental Research as a Factor in Maintaining National Security," speech to the Industrial College of the Armed Forces, November 14, 1947, AW, 31: ONR, p. 4.
8 Ibid., pp. 7, 9.
9 As quoted in Gerard, "A National Science Foundation and the Scientific Worker," p. 4.
10 See Jessica Wang, *American Science in an Age of Anxiety*, pp. 26–28.
11 Quoted in ibid., p. 31.
12 Harley Kilgore, "Science and the Government," *Science*, 102:2660, December 21, 1945, p. 633.
13 Ibid., p. 637.
14 See Bowman et al., to Truman, December 1945, DB, FA 965, 53: Federal Science Legislation, 1945–46. It is worth noting that the committee explicitly recommended jettisoning support for the social sciences, consistent with the Magnuson bill.

15 "A National Policy for Scientific Research," minutes of the American Academy of Arts and Sciences, November 14, 1945, DB, FA 965, 37: NRC Post-War Legislation (1945–46).

16 Bowman to Rabi, November 8, 1945, IRR, 65:2.

17 Committee for a National Science Foundation, "20 Leading Scientists Offer to Cooperate in Revision of Pending National Science Legislation," December 28, 1945, SRB, 2: Science Legislation Publications.

18 See Philip Powers, "A National Science Foundation?" *Science*, 104:2713, December 27, 1946.

19 Testimony of Homer Smith before the Senate Subcommittees of Commerce and Military Affairs, October 22, 1945, DB, FA 965, 37: NRC, Post-War Legislation (1945–46).

20 "Proposed National Science Foundation," *Congressional Record*, January 30, 1946, pp. 575–76. Willis was particularly enthusiastic about centralizing the role of the National Academy of Sciences.

21 Jewett to Smith, January 8, 1946, DB, FA 965, 53: Federal Science Legislation, Frank Jewett, 1945–46.

22 Julius Krug to James Webb, undated 1947, SRB, 13: NSF. Also quoted in Washington Association of Scientists, "Toward a National Science Policy?" *Science*, 106:2756, October 24, 1947, p. 386.

23 Truman, Memorandum of Disapproval, August 6, 1947, SRB, 13: NSF.

24 Dael Wolfle, "The Inter-Society Committee for a National Science Foundation: Report for 1947," *Science*, 106:2762, December 5, 1947.

25 MacKay to Bronk, June 13, 1946, DB, FA 965, 53: Federal Science Legislation, 1956.

26 Howard Meyerhof, "The Truman Veto," *Science*, 106:2750, September 12, 1947, p. 236.

27 "The Ivy-League Lobby," *New Republic*, 117, August 4, 1947, p. 10.

28 Leonard Engel, "Atomic Bombs or Atomic Plenty?" *The Nation*, 165, September 20, 1947, p. 275.

29 "Congress and Science," *The Nation*, 166:10, March 6, 1948, p. 263.

30 Engel, "Atomic Bombs or Atomic Plenty?" p. 275.

31 Higinbotham to Thurston, July 22, 1947, SRB, 13: NSF, p. 3.

32 "Science Foundation Bill," *Science News Letter*, January 22, 1949, p. 53.

33 Harold Orlans, *The National Science Foundation: A Review of the Foundation's Granting and Contracting for Research*, Brookings Institution, 1965.

34 "Foundation's First Chief," *Newsweek*, April 2, 1951, p. 52.

35 Quoted in DOD, Office of Public Information "Contributions of Dr. Alan T. Waterman to Naval Research and Development," AW, 32: OSRD.

36 Dael Wolfle, "National Science Foundation: The First Six Years," *Science*, 126:3269, August 23, 1957, p. 335.

37 See Lee DuBridge, "Policy and the Scientists," *Foreign Affairs*, April 1963, p. 581.

38 For an excellent firsthand account of service on the board, see Patrick Yancey, "My Four Years on the National Science Board," *Journal Unknown*, March 1955, DB, FA 965, 40: NSF History, 1953–69.

39 There was some discussion of the relevance of the board from the outset and certainly after 1959. One writer suggested that the 1962 restructuring had reduced the board to a "routine committee" rather than a policy-making body. See Eric Walker, "National Science Board: Its Place in National Policy," *Science*, April 28, 1967, p. 475. Also, PSAC, "Tentative Outline on Science Organization," July 14, 1961, DB, FA 965, 16: FCST.

40 Notably, both the AEC and ONR funded substantial amounts of basic research – more than did the NSF through the end of the 1960s. But the basic research funding was guided by long-term applied needs.

41 John Steelman, *Science and Public Policy*, September 1947, SRB, 13: Scientific Agencies, p. 70.
42 Robnett and Sage to Belluschi, Brooks, et. al., January 18, 1952, MIT, 155: NSF.
43 Dwight Gray, "An Interview with the Director of the National Science Foundation," *Physics Today*, 4:6, June 1951, p. 7.
44 Waterman, "Research Programs of Great Promise and Urgency in the National Interest," AW, 24: NSF, 1951.
45 It is quite remarkable that search for newer plant-based proteins was already a stipulated research goal at this time. Commercial plant-based beef substitutes would not begin to penetrate the consumer marketplace until the 2010s.
46 Waterman, "Research Programs of Great Promise and Urgency in the National Interest," p. 5.
47 See Wolfe, "National Science Foundation: The First Six Years," p. 337.
48 The number of predoctoral fellowships grew by 55 percent in the NSF's first five years, while the number of postdoctoral fellowships grew by 600 percent during this time.
49 *First Annual Report of the National Science Foundation*, GPO, 1950–51, p. 16.
50 Dwight Gray, "An Interview with the Director of the National Science Foundation," p. 8.
51 Quoted in Smith, *American Science Policy Since World War II*, p. 45.
52 "Statement of Principles for Determination of Payments to Educational Institutions Under Government Research and Development Contracts," January 30, 1947, SRB, 3: Expenditures for Research in Universities.
53 "Questions Regarding Policies of the National Science Foundation in Matters of Research Support," AW, 25: NSF 1954.
54 See Waterman, "The Role of the National Science Foundation," *AAPSS*, 327, January 1960.
55 NSF, "First Annual Report of the National Science Foundation," 1950–51, GPO, Washington, DC, p. 14.
56 Harold Orlans, *The National Science Foundation: A Review of the Foundation's Granting and Contracting for Research*, Brookings Institution, November 1965.
57 "Statement of Alan Waterman before the Committee on Appropriations (Senate)," September 19, 1951, AW, 24: NSF 1951.
58 "Federal Research and Development: Fiscal Year 1953," AW, 25: NSF 1953.
59 "Statement of Alan Waterman before the Independent Offices Subcommittee, Committee on Appropriations (House)," January 20, 1954, AW, 24: NSF 1954.
60 "US Science Department?" *Science News Letter*, January 31, 1953, p. 79.
61 NSF, "Funds for Science" (periodic report), 1952.
62 See Donald Hornig, speech at the celebration of the 15th anniversary of the NSF, 1965, DB, FA 965, 40: NSF, Loos Papers, 1951–1968.
63 Harvey Brooks, "Notes and Questions on Support of Basic Research," April 8, 1963, AW, 29: NSF, 1963, p. 4.
64 Alan Waterman, "Role of the Federal Government in Science Education," *Scientific Monthly*, 82:6, June 1956, p. 289. Lessing is quoted in NSF, "First Annual Report of the National Science Foundation," 1950–51, GPO, Washington, DC, p. 17.
65 Vannevar Bush, *Endless Horizons*, Public Affairs Press, 1946, p. 62.
66 Waterman, "Role of the Federal Government in Science Education," p. 289.
67 Ibid., p. 290.
68 Harry Kelly, "NSF Activities in Scientific Personnel and Education," September 18, 1956, AW, GC: NSF 1957, p. 7.
69 From NSF, *First Annual Report, 1950–1951*, GPO, 1951, p. 16.
70 Yancey, "My Four Years on the National Science Board."
71 Brooks, "Notes and Questions on Support of Basic Research," p. 4.
72 Alan Waterman, "The Role of the National Science Foundation," *AAAPSS*, 327, January 1960, p. 125.

73   Vannevar Bush, "Statement at the Joint Hearings of the Subcommittees of Senate Committee on Commerce and the Senate Committee on Military Affairs," *IRR*, 59:5, October 15, 1945, p. 7.

74   Brooks, "Notes and Questions on Support of Basic Research," p. 3.

75   Ibid.

76   Ibid.

77   Orlan, "A Review of the Foundation's Granting and Contracting for Research." Also, Waterman, "Notes on Federally-Supported Basic Research," AW, GC: NSF, 1957, p. 5.

78   In its inaugural year, the NSF allocated $7,500 *in total* to fellowship support.

79   Harry Kelly, "NSF Activities in Scientific and Technical Personnel and Education," September 18, 1956, AW, GC: NSF, 1957.

80   Huntress to Kispert, March 2, 1953, MIT, 155: NSF.

81   See Waterman, "Statement before Committee on Appropriations (Senate)," September 19, 1951, AW, GC: NSF, 1951.

82   During the 1950s, Princeton built the Forrestal Campus; the University of Chicago enlarged its Argonne Lab, Hopkins enlarged the Applied Physics Lab, and Berkeley opened a second campus of the Lawrence Lab at Livermore, California. All of these campuses lie apart from the home campus of their university.

83   Waterman, "Research for National Defense," *Bulletin of the Atomic Scientists*, 9:2, 1953, pp. 37–38.

84   "Federal Research and Development, 1953," RG 255, NACA, 1/IDC: Reports 1/5. The $140 million commitment was dispensed almost equally to government and industrial labs ($65 million) and university-based labs ($75 million).

85   In 1951, the Science Advisory Committee of the Office of Defense Mobilization (ODM) issued recommendations to universities receiving federal research grants on prioritizing the defense needs of the country. "everyone with scientific training and talent to use his best efforts in the most effective way to aid in defense." Science Advisory Committee (ODM), "The Role of Scientists in Defense," May 30, 1951, MIT, 194: SAC.

86   See Donald Quarles, "Speech to the Case Institute of Technology," January 12, 1954, AW, 25: NSF, 1954.

87   NSF, *First Annual Report: 1950–1951*, p. 18.

88   Quoted in DoD press release, October 6, 1953, RG 255, NACA, IDC/4: General Correspondence, 3/7, p. 1.

89   See "US Science Department?" *Science News Letter*, January 31, 1953, p. 79.

90   "Research and Development 1941–1952," MIT, 194: SAC.

91   Science Advisory Committee (ODM), "Research and the Department of Defense," July 19, 1957, MIT, 195: SAC, Basic Research, p. 2.

92   DoD press release, October 6, 1953, p. 3.

93   Rabi to Gray, July 19, 1957, MIT, SAC, 195: Basic Research, p. 2.

94   Bush reiterated this vision in a letter to Karl Compton in 1946. See Bush to Compton, December 30, 1946, VB, 609: Karl Compton.

95   See Rabi to Gray, July 19, 1957.

96   Smith, *American Science Policy Since World War II*, p. 52.

6 University Labs, National Labs, and the Culture of Science

Back to the Basics

In a 1964 speech to the American Council on Education, Donald Hornig, Special Assistant to the President for Science and Technology, noted wryly:

> I sometimes feel that Uncle Sam is considered to be a rich but not respectable relative – perhaps a bookie – who enables his nephews and nieces to enjoy a high standard of living. Everyone knows he exists, but nobody likes to admit that he matters to them.[1]

Through the 1950s, the federal government continued to grow in importance as a source of basic research funds for universities. By 1960, it accounted for 70 percent of all funding of basic research nationwide – a very rich uncle, indeed! The funding came from a variety of agencies, but over three-fourths was being earmarked for research related to either defense or health. A White House report that year reasserted that research and graduate education were the "knotted core of American science" and that the proper place for both was in the nation's universities. Strong universities undergirded a strong nation, meaning that the government's research commitment was vital to national security and strength. The interest of the nation was (in the words of the report), "that university science should be as strong as possible."[2]

That universities would be the locus of basic research had not been pre-ordained. Few universities conducted substantial amounts of research before World War II, and fewer still viewed it as a core mission. Historically, basic research had been conducted by wealthy individuals in private labs (often within their homes) or as part of a private scientific consortium of individuals through a club or society.[3] College professors through the 19th century instructed students in lab sciences and techniques irregularly; in his 1910 report to the Carnegie Foundation, "Medical Education in the United States and Canada," Abraham Flexner deemed only six medical schools (of the 155 in existence) to have adequate lab facilities and curricula to train students for the MD degree.[4] Undergraduate colleges were even less likely to possess requisite lab facilities.

DOI: 10.4324/9781003363897-6

Science education evolved rapidly through the 1940s. The great wartime successes wrought by MIT, the University of California, Columbia, the University of Chicago, Johns Hopkins, and Caltech convincingly argued that it was universities that were best equipped to shelter the brightest and most creative minds and to give them the institutional support, freedom, and time necessary for scientific discovery. Unstructured time was key, as the nature of basic research was its very lack of accountability – the somewhat meandering paths that creative minds took to happen upon discovery. No other sector seemed to have the steady revenue flows, the administrative structure, and the built infrastructure requisite for research, although small amounts were being conducted in foundation and government labs. In 1950, even before the advent of the NSF, Roswell Gibbs, the chairman of physics at Cornell and the division chairman of mathematical and physical sciences at the National Research Council, stated unequivocally, "Experience has shown that we must look to our institutions of learning to do the bulk of pure, basic, or fundamental research."[5]

The mantra of university-based basic research grew organically from the scientists already in place in research universities by the late 1940s. The professoriate was increasingly attracting scientists who craved the freedom to pursue their own projects, independent of oversight, and who viewed both government and industrial labs as inhospitable to the sort of free-wheeling research enterprise that nurtured basic science. Lee DuBridge, the president of Caltech and the wartime director of the MIT Rad Lab, emphasized repeatedly that research was an "exploration into the unknown" and therefore was difficult to evaluate by objective criteria, and others agreed.[6] The Science Advisory Committee of the Office of Defense Mobilization affirmed in a statement of basic principles in 1953 that universities must lie at the core of the nation's scientific enterprise and that government and industrial labs simply could not be reconfigured to support basic research.

Most advocates for locating basic research in universities acknowledged the synergistic roles of teaching and research. Graduate students clearly benefited from proximity to active laboratories, where they could learn proper techniques, become engaged in projects, and witness firsthand how senior scientists took a project from conception to funding to execution. It was more difficult to make the case that research strengthened *undergraduate* teaching, and nor was it easy to argue that only university-based labs could provide appropriate apprenticeships for budding scientists. The case of the reverse, that teaching strengthened and informed research, was even less evident and unpersuasive given that most successful scientists during the 1940s and 1950s used their government contracts and grants to buy their way out of teaching obligations. Graduate students could sometimes force scientists to hone their theories and research protocols through clear questions and didactic engagement, but it was difficult to make the case that this experience was necessary. A great deal of the most innovative wartime research had been conducted in the absence of graduate students, and many of the most successful scientists in government and industrial labs valued their freedom from teaching obligations. Nonetheless, no less a figure than Glenn Seaborg, Nobel laureate in physics and later Chancellor of the University of California, wrote in 1957:

> Traditionally the occupation of unfettered research has been centered in the university professor, who also has the dual role of a teacher. The union of these two activities has in my estimation been a happy one, because, on the one hand, advanced instruction is surely more inspiring if given by someone who has first-hand knowledge of the subject irrespective of his abilities as an orator, and on the other hand, there is nothing more stimulating to uninhibited thinking than the uninhibited questioning of successive generations of students.[7]

The truth was more elusive. High-quality research, both basic and applied, could be done in the absence of students, and most students (excluding graduate students performing thesis research) could ably learn from non-research-active faculty. In fact, research could actually distract faculty from investing in high-quality teaching and in the professional schools and vocational disciplines could displace the sorts of professional experiences which might be more pedagogically relevant to law, nursing, or social work students. More accurate, perhaps, was that the nation's research universities, which based their reputations in part on the accomplishments of their faculty, found that research-active faculty could enhance institutional standing and attract outstanding graduate students – two accomplishments which were only indirectly correlated with the basic work of educating undergraduates but generally not inimical to it.

Ever-increasing federal research funds to universities forced a shift in the missions of the nation's strongest research universities while also undermining traditional norms of authority and professorial practice. Funds from the DoD, the AEC, ONR, the NIH, and now the NSF accelerated a change in emphasis from teaching to scholarship, empowering individual researchers while eroding institutional loyalty. The trend had been underway in a few universities (notably MIT) since before the war, but it now came to dominate universities. In 1949, Frederick Terman, the ambitious dean of engineering at Stanford University, articulated his vision of a university as "essentially a group of scholars in intimate association with students," suggesting that learning would take place through osmosis and proximity rather than through didacticism and pedantry. In Terman's vision, top scholars through their research would attract the "cream of the crop" to their labs and research enterprises who could then be retained for local industry.[8]

This vision of the research university was deeply elitist, promoting a handful of super universities which would become nodes of scientific and doctoral training at the expense of the reputation and quality of the majority of more pedestrian schools. The view, already held by scientific leaders in the late 1940s, became mainstream through the 1950s as the most prominent scientists held increasing numbers of influential positions on science boards, advisory committees, and institutes. In 1960, the President's Science Advisory Committee (PSAC) emphasized the need to increase the number of universities harboring "superior faculties and outstanding groups of students" in an effort to advance science more rapidly. Notably, at a time when the United States hosted nearly 2,400 four-year colleges (the numbers had grown rapidly as returning GIs had sought college degrees in the postwar decades), PSAC estimated that only between 15 and 20 universities were capable of conducting serious research.[9]

Government funding distorted traditional university priorities by eroding commitments to teaching. At those universities which were most intensively focused on research, teaching diminished in importance as grant-writing and publishing (particularly in high-profile journals) rose concomitantly in status. Teaching loads "use up a man's best hours and most creative energies," wrote MIT economist Paul Samuelson to an institute colleague and needed to be sharply reduced in the interest of the nation's scientific enterprise.[10] While universities could not compete with private industry on salary, they could compete on "leisure" and research opportunity through extending summer vacations, reducing loads, and increasing support funds.[11] Glenn Seaborg went further, citing the need for the best universities to create full-time permanent research professorships, to be doled out to the most exemplary scientists who would be freed from teaching obligations permanently.[12]

The shift from teaching to grant-funded (and contract-funded) research undermined faculty loyalty to their home universities. Previously, professors had been dependent on tuition revenue to support their research projects and thus had viewed themselves as citizens of their home institution: embedded in a department, contributing to a curriculum, and often taking on a substantial portfolio of service commitments in the name of faculty governance. But generous research grants allowed traditional faculty to buy themselves a portfolio of full-time research while allowing universities to hire permanent faculty with no teaching expectations at all.[13] "Professors no longer worked *for* universities but *in* universities," wrote Harvard biologist Richard Lewontin.[14]

More generous grant-funding also eroded administrative authority. Although professors had long exercised unusual professional independence in making their work hours, controlling course content, and pursuing research questions, they had bowed to institutional pressure to serve on administrative committees and teach courses as asked. But professors who were backed by generous research grants felt less beholden to their department chairs and school deans and often withdrew from university affairs, served on fewer committees, and dictated their own modest teaching preferences. Chairpeople and deans had little leverage; faculty contracts were loosely written, and well-funded faculty could simply take their grant money and transfer elsewhere. Rebecca Lowen, in her history of Stanford University, writes:

> These traditional departmental obligations, once enforced by the head of the department, could be neglected with impunity as the individual scientist's access to patronage effectively undercut the authority of the department chair. Increasingly, access to outside funds rather than seniority conveyed power and authority within a department, and the individual professor with research contracts replaced the department as the meaningful unit within the university.[15]

Donald Hornig concurred, noting that the volatile fuel of federal grants had created a "scientific free enterprise system," stripped of loyalty and commitment to anything beyond research output and standing among disciplinary peers.

Both ONR and the NSF (and later the AEC) used a system of peer review to evaluate grant applications, similar to that used by academic journals in evaluating submissions. The system provided an important objective standard in evaluating scientific work yet unintentionally undermined the institutional integrity of universities. During this period, the peer review system began to pervade all aspects of a scientist's life, from conference participation to publication, funding, hiring, and tenure. At every juncture, the judgment of professional peers *outside* of the host institution weighed more heavily on a scientist's career than supervisory judgment *inside* the institution. Scientists correctly ascertained that internal reports counted for little in moving their careers forward and diverted their energies and attention to networking within their professional organizations and in partnering with and impressing their peers. Hornig noted that the system had moved control of departmental decision-making to outsiders with no institutional commitment – an amorphous group of professional peers who advanced or hindered careers while taking no fiscal responsibility for their decisions and judgments. He noted acerbically

> It is said . . . that the university is no longer a home where [the faculty member] feels part of the family, but a boarding house which he feels free to move into and out of, depending on the quality of the food and lodging that he gets.[16]

But although universities had few institutional tools with which to tether their star faculty members, they did possess prestige and critical pods of talented faculty which they could use to lure the next generation of promising faculty. An influx of federal funds to support both research and doctoral work might have been expected to distribute talent and research prowess more broadly across the higher education sector, but in fact the opposite was true. The most ambitious scientists tended to migrate toward the best support, facilities, and colleagues, requiring upstart universities to make massive commitments to lab construction and faculty lines to lure new talent. As early as 1945, Joseph Barker, the dean of engineering at Columbia (and later the president of the Research Corporation for Science Advancement), expressed concern that scientists who had been detailed to large-scale wartime labs would be reluctant to return to small colleges or second-tier universities. "I think it is quite possible," he wrote, "that there will be a very considerable attempted migration of research men to those institutions where research appropriations will still be available [after the war] or to commercial research laboratories like Bell Telephone Laboratory, General Electric, Du Pont, etc."[17] As the universities which had emerged from the war in the strongest position rapidly garnered the majority of federal contracts and grants in the 1950s, the prognosis proved to be true.

Part of the problem was the "indirect" funding built into federal grants and contracts. At base, a research grant or contract reimbursed the host institution for the researcher's salary and benefits on a prorated basis, as well as any new equipment or materials which would need to be purchased to carry on the research. (In theory the university could recover the salvage value in the equipment at the conclusion of the research, but in fact that value was rarely charged against the contract.) These costs were known as "direct" costs and were beyond dispute. But the universities bore a large array of costs related to supporting a research enterprise though not attributable to any particular research project. These costs, known as "indirects," included administrators' salaries, depreciation on lab buildings, partial use of the library, general maintenance and upkeep of related buildings, and utilities.

During World War II, MIT had negotiated a flat 7 percent indirect rate on all military contracts which it found just barely adequate to cover costs associated with the Rad Lab and other wartime endeavors. Faced with the reduction of this indirect benefit to 3 percent at war's end, the institute argued strenuously that such a move would force it to divert resources from other activities (notably teaching) to subsidize the government and would thus impede cooperative ventures between the military and the university.[18] The university prevailed.

Rapid escalation of government contracting after the war, largely through ONR and the AEC but through other agencies as well, forced more universities to grapple with the true costs of taking on contracted research. Gradually, they persuaded government funders that the true cost to the university was substantially higher than the direct cost budget might indicate, and that rather than argue for prorated depreciation on infrastructure they would settle on a substantially higher flat indirect rate to be negotiated with each university.[19] PSAC recommendations in 1960 suggested removing the cap entirely and simply reimbursing the universities the "full cost of research performed for the government – including overhead."[20] By that year, as well, the government acknowledged that the nation's research universities probably needed an investment of $500 million for capital construction to be able to adequately perform the government research (and associated doctoral training) which would be demanded of them over the next decade.[21] In other words, the existing 24 percent indirect rate had been inadequate to fairly reimburse the universities for the tasks demanded of them.

The issue of indirects was far from simple and reflected the complex interrelationships between teaching and research, particularly at the graduate level. Viewed reductively, professors could spend their time either teaching or performing research, and reimbursement for the two separate tasks needed to come from two clearly distinct budgets: tuition revenue and grant support. But if teaching and research were, as claimed, *complementary* activities, then outside support for research effectively subsidized teaching, just as tuition revenue meant to pay for teaching was indirectly supporting research. (The first scenario was generally held to be more compelling.) Terman, of Stanford, in repeatedly calling for only the very best researchers to staff the university's engineering faculty after the war, was clear that a doctoral program devoid of active research opportunities was pointless, and leaders of most of the nation's top research universities agreed.[22] In the 1960s, Donald Hornig went as far as to dismiss any university which could not sustain a top-flight research program as weak and incapable of supporting excellent teaching.[23] Seen in this light, conducting research for the federal government was not so much an onus to be borne by universities but rather a critical input to their core mission, without which they could sustain neither their academic programs nor their reputations.

Fellowship funds made available by the National Defense Education Act in 1957, further weakened university claims of the great costs they were incurring by performing research for the government. Fellowships made available at that time, when added to previously available fellowships, allowed the government to fund 17,000 graduate students by 1962, which included 29 percent of engineering students, 37 percent of physical science students, 46 percent of students in the life sciences, and 39 percent of students in the social sciences. It made matching funds available for dormitory construction and nearly $460 million annually for laboratory and classroom construction.[24]

By 1960, the extraordinary level of federal support for graduate education – both from research support of the labs in which graduate students would perform their doctoral investigations and from fellowships to the students to reimburse them for tuition expenses and basic living needs – undercut university complaints about the "soft costs" of grant-funded research. In fact, the opposite appeared true – that federal contracts and grants were highly desirable as they allowed universities to enlarge their faculties with high-salaried research professors. Such faculty stars, in turn, attracted the best doctoral students and increased university prestige while costing the university very little. Grant-funded laboratory buildings, cyclotrons, synchrotrons, electron microscopes, telescopes, and centrifuges raised universities in national rankings of academic programs – which ultimately served to increase alumni loyalty (and donations) while drawing the strongest under-graduate applicants. The benefits of federal science funds were broad indeed and the costs nominal.

Except, there *were* costs, but they were more philosophical than fiscal. By the mid-1950s, with the NSF up and running, 60 percent of all federal funding to universities was still coming from the defense-related agencies (ONR, DoD, NACA, and the AEC), with most of the remaining funds coming from the NIH. There was no conceivable way that the NSF could grow fast enough to displace this substantial funding stream, and budget-ing trends indicated that, if anything, the role of the DoD in basic research would continue to grow. The NSF was not designed for the sort of massive, multiyear, multi-institutional grants that funded huge pieces of scientific apparatus (like accelerators and reactors) and drew on the work of dozens of scientists. DuBridge noted that Caltech would "go broke very promptly" if all funding of basic research were suddenly transferred to the NSF.[25]

But scientific funders demanded some level of accountability, and through the 1950s, defense-related agencies impelled scientists to bias their work toward defense interests. The pressure was more a tug than a push. The DoD and the AEC were hardly forcing scientists to work for them, but as the source of the preponderance of research funds they lured scientists in the engineering and physical scientists to focus their work on defense-related problems. Senior defense executives even deemed the trend desirable and imposed on universities an obligation to "make the best use of the scientific strength of their establishments to strengthen the nation's defense," in the words of one PSAC memo.[26] The DoD, the nation's single largest funder of university research (both basic and applied) became more aggressive on this point through the decade, viewing university science departments not so much as potential defense resources but rather as vital national security assets which should be placing themselves at the service of the military. PSAC wrote in 1951, "It is our general point of view that at this time a substantial portion of the nation's scientific effort needs to be diverted to defense programs, surely no less than 20 percent and probably more."[27] Else-where, PSAC queried, "What can the Science Advisory Committee suggest to an individual scientist to enable him to use his efforts more effectively for national defense?"[28] Possible moves included detailing university faculty to DoD internal labs or to scientific administrative posts in the Pentagon; greater commitment to teaching release on the part of academic administra-tors; and more flexibility in moving scientists temporarily between institu-tions so as to better leverage their skills in service to a larger working group.[29]

Funding pressure insidiously pushed young scientists to take on research agendas oriented toward military applications, even when the research itself was more fundamental. As research universities hired greater numbers of faculty on predominantly research lines (meaning that there wasn't ade-quate tuition revenue to fund the positions), young scientists were made to understand that they would be largely responsible for paying themselves out of grants and contracts and thus needed to take on projects with the best chance of being funded. This was hardly the traditional role of university-based basic research, which was supposed to orient itself toward fundamental questions regardless of their applicability or relevance. In theory, this was supposed to be the core distinction between university science and industrial (or government) science, where the industrial labs busied themselves with solving specific problems *en route* to producing a new marketable product or weapon while the university labs performed the innovative theoretical work which might lead to future applications. A 1953 internal working memo from the RDB explicitly described the preferred relationship between the DoD and a contracting university lab as follows:

> [T]he scientists be legitimately encouraged to be interested in fields which are of potential importance with respect to the national defense, so that the entire scientific strength of the country could be brought to bear promptly and effectively in case of a severe emergency, so that the Services are continuously and growingly aware of scientific developments and of the value to them of scientific activity, and so that the scientists and the research administrators can contribute an important element of intellectual leadership within the Armed Services. . . . As a by-product of great importance, this support of science tends to assure to the Armed Services a more adequate supply of suitably trained scientists who may later find employment within the Services.[30]

An adjunct concern to the relentless pressure to push university science and engineering toward a military posture was the added pressure of security and classification. Science, by necessity a deeply open and transparent enterprise, was bounded by security considerations when funded by the security services. Scientists needed to gain permission to publish results or even share data with colleagues. Scientists with politically marginal pasts – membership in socialist or communist parties or even simple association with communist organizations – could find themselves excluded from funding, regardless of their past service and oaths of loyalty. The measures, which had started quite reasonably during World War II, had morphed into a culture of secrecy under the threat of Soviet espionage. While Vannevar Bush had urged a reasonable balance between wartime secrecy and scientific transparency, Cold War culture pushed the balance toward near paranoia.[31] ONR, the AEC, and the RDB operated under codes of strict secrecy, but even the NSF could not escape the prevailing zeitgeist. In 1954, the National Science Board endorsed a statement that the Foundation did not "knowingly give or continue a grant in support of an avowed Communist or an individual convicted of sabotage or other crimes involved in the nation's security."[32] While the standard of "convicted sabotage" was reasonable, the policies tended to have a chilling effect on the scientific community, calling into question youthful engagement with leftist politics or even engagement with free thought.

Working at the Federal Labs

In parallel to the extramural grant and contract programs, the federal government continued to fund substantial amounts of basic and applied research in the physical sciences in its own labs. Of the many federal installations, three in particular were designated "national laboratories" after the war – Argonne, Oak Ridge (Clinton), and Brookhaven – and several others operated essentially as national labs, including the Radiation Laboratory at the University of California (later named Lawrence Berkeley Laboratory) and the Los Alamos complex, which was also operated by the University of California. All of the labs, with the exception of Brookhaven, were successor institutions to wartime locales of the MED.[33] Each grew under renewed funding commitments during the decades that followed. By 1958, the combined budget for the national labs was $206.3 million, and fully $8 billion had been invested in the national labs on both construction and operations since the war.[34] When combined with government spending in non-AEC internal labs (largely the Naval Research Lab and the labs of the NIH), the total budget of federal labs by the end of the 1950s constituted 30 percent of all spending, nationally, on basic and applied research.[35]

Organizationally, the national labs were odd creatures. Henry Smyth, a member of the AEC, asked an assembled audience in 1949 to acknowledge that the whole scheme was, "frankly . . . a novel one."[36] The labs were owned by the federal government but delegated to an assortment of private and public universities, university consortia, and private companies for management and oversight: Argonne to the University of Chicago, Lawrence Berkeley to the University of California, Brookhaven to a tight consortia of nine private northeastern universities, and Oak Ridge to a revolving cast of university and corporate supervisors including the University of Chicago, Monsanto, Union Carbide, and briefly to a consortium of southern universities known as the Oak Ridge Institute of Nuclear Studies.[37]

The national labs, somewhat distinct from the Naval Research Lab and the National Institutes of Health, existed in part to aggregate a critical mass of research talent in one place to work on classified projects, but more importantly because the cost of research in the nuclear realm was simply beyond that of any one university. Reactors and accelerators were enormously expensive to build and maintain and also posed substantial security issues. The national lab system allowed the government to fund atomic research at levels beyond what it could reasonably delegate to any one university while granting scientific oversight to university partners which were more attuned to the milieu of science and at the same time maintain control over security and secrecy. As an added bonus, universities could assign graduate students to the laboratories to fund their studies while also giving them access to innovative research on which to base their doctoral research.[38] Not truly national, they functioned more as "regional hubs" for scientists wishing to work with reactors and accelerators.[39]

Several additional non-AEC laboratories grew during the postwar years, each built on similar government–university partnerships. The Applied Physics Laboratory (APL), a joint venture between the Carnegie Institution's Department of Terrestrial Magnetism and Johns Hopkins University's physics department developed the proximity fuse during World War II. The venture, initially coordinated by the OSRD, grew rapidly; by the end of World War II it employed nearly 200 physicists and engineers. The Navy considered the fuses so vital to the war effort that it transferred $2 million to the project, making the APL "more a ward of the Navy than of the OSRD," in the judgment of historian Michael Dennis.[40] Hopkins was initially leery of the relationship, which violated university norms of independence and openness but was ultimately won over by the sheer volume of funds that the Navy was willing to invest in the venture.

Careful negotiations involving the university's president, Isaiah Bowman, and the director of the lab, professor of physics Merle Tuve, produced a tighter, more permanent relationship after the war, in which the university agreed to manage the lab and hire research personnel but not to create permanent faculty lines for researchers associated with the lab. In so doing, the university gained a hugely wealthy partner institution in which it could place faculty and students for research without actually owning the work of the lab or controlling its agenda. The work of the lab was devoted almost wholly to military ends, which could potentially threaten the university's commitment to academic independence should the relationship become overly intimate. Nevertheless, by 1948, Bowman agreed to elevate the lab to a true division of the university, on par with the medical school, the graduate school, and the undergraduate liberal arts college.

The Jet Propulsion Laboratory (JPL), affiliated with Caltech, evolved similarly. Established by the Army in 1943 (grounded in an existing university aeronautical engineering lab), the lab developed ballistic missiles during the war and teamed with Wernher von Braun's Army Missile Lab during the 1950s to work on the Redstone rocket. In 1958, it was transferred to the newly created NASA. As was true with APL, JPL had an uneasy relationship with its partner university, which viewed the classified military research as inimical to its core scientific mission. Nonetheless, the funding was simply too generous for the university to lightly surrender the relationship. The lab allowed Caltech to become a leader in aeronautical (and later aerospace) engineering while also helping to attract top faculty.

Both APL and JPL proved troublesome to their host universities for the secrecy that surrounded their work, their near-exclusive focus on military projects, and the highly applied nature of their research. (Both labs devoted less than 5 percent of their research budgets to basic research.) Both were located some distance from campus –JPL was eventually moved out of Pasadena entirely, and APL was located in Silver Spring, Maryland, closer to Washington than to Baltimore. Neither lab had the ability to create permanent faculty posts with tenure.[41] Both, however, played host to faculty and graduate students' research projects and strengthened university ties to regional industrial firms; many aerospace companies in southern California, for example, maintained research ties to JPL. At the same time, both labs proved to be administratively troublesome to the military, which resisted their independence and lack of clear accountability – historian Roger Geiger noted that JPL was a source of "intermittent irritation" to the Army.[42]

After some hesitation, both universities committed to the relationships. Patriotism played a part, no doubt. The Soviet threat was compelling and present, and universities feared appearing overly aloof amid Cold War concerns. But the benefit to the universities' academic mission was the true driving force; the labs were simply too valuable to lose. In both cases, the presence of the labs greatly boosted the host institution's profile in an important field of engineering, allowing its faculty and students to engage in projects which would have simply been unavailable in the labs' absence. The question was one of managing the relationship – exploiting the wealth of opportunity without succumbing to a culture of secrecy and militarism. Ultimately, both Caltech and Johns Hopkins were forced to grow more comfortable with the relationship, embracing more "porous boundaries separating the civilian from the military."[43]

One of the great advantages to the military of partnering with the universities was the ability to hire and retain research talent above government scale. Government scientists were embedded in the GS pay scale, with the most senior scientists (those managing a large laboratory) holding the GS–15 pay rank. This rank reflected private sector managerial salaries, but did not account for substantially higher salaries of research managers in either the private or academic sectors. All senior government employees were losing purchasing power as private sector salaries (and prices) increased after the war. By 1951, one internal government estimate placed the GS 15 salary at only 71 percent of its purchasing power of 1939.[44] The problem only grew worse in ensuing years with one interagency panel in 1957 reporting challenges in both recruiting junior researchers and retaining senior researchers.[45] Researchers hired on university lines could draw salaries above the federal scale, allowing the government a work-around to bring in necessary talent to staff its labs.[46]

The government, itself, was a major source of salary inflation. In funding so many engineers and scientists through the 1950s, in both extramural and intramural research, the government had created a nationwide shortage of trained research scientists and engineers for which universities and industrial labs competed. Moreover, the allure of government work was fading for researchers generally, who complained of a culture of "unimaginative plodding" as described by one reporter.[47] One internal report cited the many reasons that scientists left federal employment, including "dissatisfaction with red tape . . . slowness of advancement . . . freedom from annoyances by administrative staff."[48] University management of federal labs assuaged resentful scientists to some degree, granting them the prestige of university affiliation, higher salaries, and fewer bureaucratic barriers. The arrangement

seemed to include the best of both worlds – the security of continued government funding and the freedom of university life, so long as the relationship held. Through the 1950s, the relationships only grew stronger.[49]

The labs themselves wrestled with achieving the correct balance between
civilian and military orientation, granting substantial authority to their partnering universities while assigning a military liaison. Assigned officers vac-
illated in their level of engagement and supervision, demanding that the
labs stay focused on their military mission while trying not to squelch the
creative impulse required of scientific inquiry. The balance required a dance
of sorts between military and civilian leadership, with academic directors
pushing back, as necessary, when military supervision became overly tight.
Henry Smith, one of the AEC's commissioners, noted,

> There will probably always be a tendency for those who pay the bills to
> ask for too strict an accounting. If that happens, I think it is the business
> of the laboratory directors to remind . . . Washington . . . that research
> cannot be run that way.[50]

Part of the problem lay in the military's system of rotating billets for officers.
Officers in charge of military oversight for the labs often rotated every two
to three years, denying them the chance to really familiarize themselves with
the work of the lab and precluding the creation of tight, working relationships with the senior scientists directing major research projects. The military
insisted on viewing the labs as ordinary military posts rather than the unique
assignments that they were. A frustrated Lee DuBridge wrote to the chair of
the House Subcommittee on Military Operations that these frequent rota-
tions ultimately undermined the military–civilian relationship. He wrote,

> I do not mean that there are no military officers who can be superb
> laboratory directors. I do mean that even if a particular laboratory has a
> superb director for two or three years, he may then be transferred to some
> other duty and the continuity will be broken. As long as the position of
> the director of a laboratory is regarded as a military post, subject to regu-
> lar rotation, continuously good leadership cannot possibly be achieved.[51]

Postwar MIT

By the end of World War II, MIT led the nation's universities by a considerable margin in the size and scale of its military contracts, with 75 contracts
worth $117 million, ahead of second-place Caltech with $83 million and
third-place Harvard with $31 million.[52] Over the ensuing two decades, MIT
would maintain the scale of its military contracts while altering the structure
of its contracting terms. Free-standing divisions like the Rad Lab would
morph into interdisciplinary shops answering to multiple academic departments. At the same time, the university created several research centers such

as Lincoln Laboratories and the Instrumentation Laboratory – using a model similar to the national labs, JPL and APL – wherein the university was contracted to manage and staff a federally owned facility. So substantial was the scale of military research at MIT that by 1962 Alvin Weinbert, the director of Oak Ridge National Lab, joked that it was difficult to tell "whether the Massachusetts Institute of Technology is a university with many government research laboratories appended to it or a cluster of government research laboratories with a very good educational institution attached to it."[53]

The first order of business was to hold on to military (particularly Navy) contracts by enlarging faculty in areas of key interest. During the war, MIT had loaned a number of its faculty to the Rad Lab and the Manhattan Project, and it now sought them back. At the same time, it began negotiating with the Navy for greater long-term commitments to several key areas, including nuclear propulsion, nuclear armaments, and instrumentation. In Fall 1945, John Slater, the chairman of the physics department, reached out to the Navy to assure them of the institute's growing commitment to high-energy nuclear physics and engineering – the fundamental work necessary to future exploitation of atomic power.[54] At the same time, the institute began to aggressively recruit new faculty and to woo back faculty who had been detailed to various government and industrial labs during the war, such as Jerrold Zacharias, Bruno Rossi, and Victor Weisskopf from Los Alamos; Charles Coryell from UCLA and Oak Ridge; and I. A. Getting, who would take charge of the university's synchrotron.

Faced with the challenge of melding the work done in the Rad Lab and the labs of the MED with the university's core academic mission, MIT in early 1946 created the Laboratory for Nuclear Science and Engineering (LNSE) to conduct work in high-energy physics, fission, and particle behavior. The new lab would answer to the departments of physics, chemistry, electrical engineering, chemical engineering, mechanical engineering, and metallurgy and would function both as a locus of government contracted work and as a training facility for MIT students.[55] LNSE differed from the Rad Lab insofar as it reported to academic departments rather than being wholly independent of them, but the scale of its work would require substantial grants and contracts from the military all the same. Although the university would never manage its own reactor (as did the Universities of California and Chicago), it would rapidly acquire a large accelerator and become one of the founding members of the consortium of universities, which operated the Brookhaven National Laboratory on Long Island. All of this was dependent on a steady stream of military contracts, and thus, as early as 1946, an internal working group stated that to get the requisite personnel, material, equipment, and classification to do innovative nuclear research, "we need a contract with the Army, the Navy, or preferably with the Atomic Energy Commission."[56] Within two years, the university would have government contracts for nuclear research totaling nearly $3 million, with the Navy doubling its commitment each year.[57]

In the realm of particle physics, MIT's most important decision during the early postwar years was to acquire a new, higher-energy accelerator. After dropping the idea of purchasing a betatron from General Electric, it designed and built its own 300-Mev synchrotron over the next five years at a 17,000 square-foot site just off of the main campus. The synchrotron (described as a "giant slingshot" in the university's press release) accelerated electrons to speeds great enough to produce mesons – subatomic particles which seemed to explain the strong forces which bound together the nucleus of an atom. Over the following years, MIT researchers would use the machine to measure meson decay times, bow-wave radiation, X-ray degradation, and the interaction of mesons with gas nuclei in a cloud chamber.[58]

The institute performed a neat administrative pivot when it transformed the Rad Lab into the new Research Laboratory in Electronics (RLE). Again, the goal was to expropriate a government-owned research effort and bring it under the mantle and oversight of academic departments. Like LNSE, RLE was overseen by a consortium led by the electrical engineering and physics departments, and like the national labs it came into existence with the considerable patrimony of wartime equipment purchased by the military. RLE also resembled LNSE insofar as it was an interdisciplinary lab, funded from outside sources (primarily the Navy) which would host basic research conducted by doctoral and postdoctoral researchers under the leadership of MIT faculty members. Initial areas of interest, besides a continuation of the Rad Lab's work on microwave radar, included the properties of gaseous conduction, high-frequency electromagnetic fields and acoustic waves, and digital calculation (protocomputing).[59] In its first five years, the lab won large contracts with the Navy to design the guidance system for the Meteor missile. At the outbreak of the Korean War, the Navy doubled its contracts with RLE.[60]

RLE also helped MIT forge new relationships with industrial firms. In the late 1940s, MIT developed close working collaborations with Sperry Gyroscope, Sylvania, and Raytheon – three firms which piggybacked on the military contracts with RLE and regularly hired graduates out of the institute's electrical engineering program. Bell Labs began to contract with RLE in 1951 on digital communications networks, committing $25,000 per year for work on neural relays, and by 1953, over half of its transistor work was supported by military contracts.[61] Likewise, Raytheon grew rapidly through government contracts in the 1950s. In a decade its military work grew tenfold; military funding accounted for 88 percent of the Raytheon's revenue by 1960.[62] At the same time, a number of faculty and graduate students associated with the lab started their own companies during the 1950s, most specializing in microwave electronics, digital computing, and radar components.

These tight relationships with both the military and the industry were beneficial to the scope and quality of the university's research effort but inevitably biased faculty interest toward military applications. One MIT professor who ultimately decamped to Swarthmore College observed, "Professors teach what they know. They write textbooks about what they teach. What they know that's new comes mainly from their own research. It is hardly surprising, then, that military research in the university leads to military-centered undergraduate curricula."[63] So great was the military influence on the institute over the following decade that a 1956 faculty committee suggested that military contract research was undermining the institute's teaching mission. The report read:

> Too great a proportion of our research funds in engineering involves the purchase-order type of commitment which has characterized so much of postwar research. Often the result is poor research, or something which is not research at all. Moreover, the very spirit of such an arrangement can easily pervert the educational role of research at a university.[64]

Other MIT labs replicated the model and the success of RLE. Charles Stark Draper's wartime instruments lab, where he had conducted work on gyroscopes for Sperry and the Navy, morphed into the postwar I-Lab, which worked with the Navy on gun and bomb-sights and with the Army on aviation navigation instruments.[65] At a lesser scale, Jay Forrester took charge of the Servomechanisms Laboratory, which worked on flight simulators during and after the war, as well as on avionic wind simulators and aeronautical design. RLE, Draper Lab, and the Servomechanisms Lab all worked closely with industry, subcontracting out portions of their military work to local firms which, in turn, took advantage of the stream of capable graduate students moving through MIT's graduate programs in electronic, aeronautical, and mechanical engineering. The model of piggybacking graduate programs on contract research for the military was not unique to MIT but certainly most pronounced there. James Killian, president of MIT in the 1950s, described Draper's success: "Draper thus taught engineering by *doing* engineering; learning and research were intermingled, and the students' learning experience became memorably provocative and exhilarating. The art he brought to teaching was another of the advances in MIT education in the 1950s."[66]

Military contracting with MIT reached its apex with the creation of the Lincoln Laboratory in 1950, MIT's counterpart to the JPL at Caltech and the APL at Johns Hopkins. Lincoln Lab was created by request of the Air Force, which wanted an MIT-affiliated lab where it could conduct classified research on air defense radar and antiaircraft weapons. In 1951, General Hoyt Vandenberg, the Air Force chief of staff, requested that MIT work with the Air Force on a new venture built on previous work in the Instruments Lab and RLE but under tight classification and with total commitment to Air Force defense problems. Although several scientists at RLE feared that the new venture would cannibalize existing MIT labs, the Air Force's generous funding commitment, coupled with the heightened pressure of the Korean conflict, impelled the lab's creation. Julius Stratton, a professor of physics (and later MIT's chancellor) wrote, "we came to the conclusion that the threat of war had now become real and serious and the institute owed a major duty to the country."[67]

Ultimately, Lincoln Lab absorbed many of the faculty and projects previously housed at RLE while continuing RLE's practice of subcontracting with industrial firms. As the lab expanded, its space requirements necessitated a move to suburban Boston, where it could establish greater independence from its parent academic departments while continuing to host faculty and graduate student research. And as electrical and aerospace engineering grew increasingly complex as disciplines, the substantial resources of the lab facilitated MIT's hold on the top tier of those disciplines. One of its directors would point out in later years:

> What . . . can Lincoln Laboratory do for engineering education? It can provide entirely new opportunities to advanced students who wish to work in the complex fields at today's technological frontier. . . . It is now becoming apparent that graduate education in many of today's crucial problems requires the resources of research centers larger than MIT's laboratories in Cambridge. In radar and in space surveillance, in radio physics and astronomy, in information processing and in communications, major physical facilities are available at Lincoln Laboratory. There is a growing recognition among educators that these facilities are not regrettable manifestations of mid-century complexity, but are the environment in which tomorrow's scientists and engineers must learn to live and work.[68]

Through the 1950s and 1960s, MIT excelled at partnering with the Navy and the Air Force on research and training. The relationship was symbiotic, with the Navy frequently exploiting the preeminence of MIT faculty in the disciplines it found most relevant while MIT could expand its footprint and its faculty on the substantial funds coming in from its military partners. As the MIT faculty grew ever stronger during these decades, the university's bonds with the Navy and the Air Force grew stronger still, even as it leveraged those relationships to recruit the nation's best graduate students to its programs in electrical and aeronautical engineering and to physics and physical chemistry – now all regarded as the best in the nation. So tight was the relationship that the Navy even invited MIT to partner with it on strategic planning. In 1955, the Chief of Naval Operations invited President Killian to appoint select MIT faculty to a long-range planning group in recognition of the central role that the institute's scientists had played, and were continuing to play, in the evolution of naval tactics, navigation, radar, and air defense.[69] Although some faculty expressed concern at the influence which military problem-solving was having on the nature of education at MIT, most faculty and administrators recognized the relationship as an extraordinary opportunity for faculty and students alike, provided that all were willing to orient their interests toward military need.

Keeping Up With MIT

MIT was first among American universities in the scale of its military and AEC contracting but hardly unique. The University of California continued to manage the Berkeley and Los Alamos labs, as well as its own Radiation Lab, as did the University of Chicago at Argonne, Caltech at JPL, and Johns Hopkins with APL. Princeton acquired an 800-acre tract of land ten miles from its campus in 1950 (called, at the time, the "second Louisiana Purchase") on which it would construct its vast Forrestal campus devoted initially to naval research.[70] Columbia, Cornell, and Harvard, first, and later the Universities of Illinois, Michigan, and Washington all developed sizable contracting relationships with the armed forces and the AEC.

No university set its sights on replicating MIT's success more aggressively than did Stanford. Frederick Terman, a professor of electrical engineering who had been detailed to the Radio Research Laboratory at Harvard during World War II, became the visionary of modern Stanford in his roles as dean of engineering and provost during the 1950s and 1960s. Envious of Harvard's research prowess and prestige, he recognized that Stanford would need to follow MIT's model of building up contract research with the armed forces and the AEC to lift itself to the first rank of research universities. Eschewing undergraduate programs as irrelevant to a university's standing, and focusing on a small number of disciplines in the engineering and physical sciences, he promoted the "steeple concept" of building a few extraordinary programs rather than many programs which were simply very good. And all would be predicated on the university attracting and retaining some of the best faculty researchers in the world in disciplines that could attract funds and influence the course of industry.[71]

Under Terman's leadership, Stanford's research output advanced rapidly. In 1946, its government contracts totaled only $127,000; 13 years later they were at $13 million. During this time, Terman encouraged Stanford to replicate MIT's model of partnering closely with industry both as a source of research funds and also as a professional destination for recent graduates who could help to build relationships between the university's faculty and industrial researchers. Through the 1950s, Stanford developed close relationships with Bell Telephone, Pacific Telephone and Telegraph, General Electric, RCA, Zenith, and Raytheon. Also, it guaranteed generous funding to graduate students, allowing it to build one of the nation's top three electrical engineering graduate programs by 1960.[72] As was true with many universities, Stanford flourished during the Korean War, during which the armed forces doubled their contracting commitments to the university with the bulk of the funds going to two semiautonomous labs: the Applied Electronics Laboratory and the Electronics Research Laboratory. As Stanford graduates streamed out of these highly touted programs and labs, they founded or helped to build multiple firms in California, including Varian Associates (developer of sophisticated vacuum tubes), Litton Industries (navigation and electronic warfare equipment), Eitel-McCullough (transmitting tubes), and Watkins-Johnson (solid state devices and electronic warfare subsystems). These firms, in turn, partnered with Stanford faculty and hired Stanford students and graduates.[73]

Stanford cemented its position as the foremost locus for engineering and applied research on the West Coast when it attracted a new 30-Bev (billion electron volts), two-mile-long linear accelerator in 1960 (SLAC) on the recommendation of a special White House advisory committee chaired by Emanuel Piore, the director of research at IBM. The committee acknowledged the need for a new and larger accelerator to be regarded as a "national facility" and deemed Stanford to have the requisite space and faculty resources to support the project.[74] The total construction cost of $100 million could not be credited to Stanford's contracting account per se, but it was an indicator of the centrality of Stanford by that time to the nation's military and particle physics research effort. At the time of its construction, SLAC was the world's longest linear accelerator, the country's longest building, and the world's straightest object.

In later decades, the MIT/Stanford model of government contracting would become more commonplace, and other universities sought to build research prowess and prestige on a steady flow of DoD and AEC funds. In each case, however, the relationships diverted institutional energy and attention to projects of particular interest to the military. Conducting DoD research was not so much morally problematic as it was possibly corrosive to a university's fundamental orientation.[75] Outside funding tended to draw faculty attention to areas of interest to the funder rather than to more fundamental questions within the discipline itself. However, to refuse such funds relegated a research university to second-class status. The distortion threatened to undermine universities' academic independence and compromise their ethos of independent thought and free exchange of ideas.

Particle Physics at the Center

Although fission and reactor development had grabbed the spotlight during the war years, through the 1950s the federal government's focus swerved toward particle physics. The general atomic model had evolved rapidly through the 1940s toward a more complex understanding of the subatomic particles which made up the traditional atomic troika of proton, neutron, and electron. These new particles included the neutrino, the meson, the pion ("pi-meson"), and the strange particles. All existed at the threshold of wave–particle duality in which energy and matter seemed to meld into a continuum, and all were far too small to be "seen" in any conventional sense through reflecting light waves within the visible spectrum. Moreover, many could only be separated out after a neutron or a proton was struck by a counterpart whose energy state had been elevated to millions of times its ordinary resting state and even then might only exist independently for a very short time – perhaps a millionth of a second.

Neutrinos were the most common of these new particles but also the smallest, lightest, and hardest to detect. They were emitted by radioactive nuclei whenever electrons were emitted and like light waves had no mass at all and apparently no interactions with other forms of matter. Neutrons traveled so seamlessly that they could easily transit through the Earth. The hundreds of billions of neutrinos emitted by the sun each second *per square inch* were as likely to pass through a person whether daytime or nighttime, the only difference being that at nighttime the neutrinos had first traveled through the Earth from the sunny side.

Muons were similar to electrons in mass and energy and existed for only a millionth of a second after being emitted from an atom, thereupon decaying to an electron and a neutrino (and an antineutrino). Pions, a bit heavier than electrons, seemed to play an important role in binding the protons and neutrons together in an atom's nucleus – a sort of atomic glue. Protons and neutrons exchanged pions constantly while bound in the nucleus. When the pions were freed, however, they rapidly decayed into muons, neutrinos, and photons (elementary bundles of light energy) in roughly one-millionth of one-billionth of a second. The strange particles (later called quarks) were only just being theorized in 1960; they would later become the basis of a broader and more robust model of subatomic (hadron) formation.[76]

The realm of particle physics became the most exciting area of physics through the late 1940s and 1950s, and physicists sought to understand the fundamental interactions at the base of matter and create a unified theory of the forces of the universe. The path forward was in building a series of ever more powerful particle accelerators in which protons or neutrons could be accelerated to extraordinarily high speeds, smashed into other particles, and thus peel off subatomic particles for extremely brief periods of time which could be observed *en masse* through their radioactive footprints. While Lawrence's original circular accelerator, the cyclotron, had been less than a foot in diameter, the proposed newer accelerators were many times that size with the capability of raising the energy in the accelerated particles to hundreds of billions of electron volts. The new machines were enormously expensive to build and operate – far beyond the budgetary capabilities of any one university – and all were financed by either the AEC or ONR (with the Air Force's Office of Scientific Research stepping in starting in 1954). From 1946 to 1958, the AEC's budget for accelerator work rose from $4 million to $28 million and was slated to rise to $70 million by 1963 (and possibly $600 million by 1973).[77] The country had 15 major accelerators by then (at 200 Mev or greater), with four more under consideration (at Madison, Rochester, Oak Ridge, and Stanford).[78] The new accelerators would be capable of boosting particle energy to billions, rather than millions, of electron volts and were slated to cost over $100 million each in construction costs. The annual operating cost of just the four new accelerators would be between $15 million and $20 million each, and the estimated operating cost for just *one* experiment on the SLAC (then under construction) was $850,000.[79]

As usual, fear of Soviet supremacy drove some of the planning, with an ominous 1958 warning ("leadership is now being challenged by the Soviet Union") impelling a White House planning group to pressure the Pentagon to redouble its commitment to the field.[80] An NSF advisory panel on higher-energy accelerator physics explicitly directed the federal government to continue its active support of the field (including the design and construction of newer and larger machines) and the DoD, the NSF, and the AEC to coordinate their efforts to provide a super-funding stream to maintain the existing facilities while increasing the operating budget for future experiments.[81]

Accelerators, even more so than reactors, played at the interstices of the appropriate military role in funding basic research. The machines were enormously expensive to build and operate and getting more expensive as scientists required ever-larger machines to advance the state of the science. (The trend would reach its apotheosis with the planned, though never-built, Superconducting Super Collider – a 54-mile ring to be built in Waxahachie, Texas which would elevate protons to 20 *trillion* electron volts, at a cost of $4.5 billion.) The work on the machines was science at its most fundamental – seeking to answer the most basic questions about the nature of matter and energy. But it was easy to see how such research might produce weapons applications in the future, given the military's focus on producing more powerful and more efficient nuclear weapons. The DoD and the heavily arms-oriented AEC felt that such work lay squarely in their purview and enthusiastically funded the ever-growing accelerators.

At a dedication ceremony of the Princeton-Pennsylvania accelerator in 1963, Glenn Seaborg, the chairman of the AEC, noted the shared nature of the project which welcomed all qualified scientists, regardless of institutional affiliation, to exploit the research potential of the new facility while serving the interest of the public. He noted that the government would need to justify such enormous expenditures to both the scientific community and a "large and interested lay public."[82] Certainly, such work lay at the forefront of scientific knowledge and was a critical component both to advancing science and to training the next generation of physicists. But the question of whether or not the public would tolerate the investment of such vast sums, short of potential practical applications, was become increasingly pressing. It was one thing to invest billions of dollars in beating the enemy and quite another to invest those funds in refining theoretical models of particles so miniscule and ephemeral that they could never be seen and existed for only picoseconds before disappearing into the atomic matrix.

The paradox can be resolved, however, by viewing particle research at the time as foundational work in advancing general knowledge of the nature of the atomic nucleus. Given the importance of nuclear physics in developing new generations of fission and fusion weapons, the AEC could justify its substantial investments in accelerator development and experimentation as more relevant to national security than might be apparent to our contemporary eyes. Particle experimentation of the 1950s was the empirical side of post-Newtonian physics – the practical measurements which would flesh out the practical ways in which fundamental forces and particles interacted, morphed, combined, or split off. While Congress would eventually refuse to fund ever-more expensive accelerators (notably the Superconducting Supercollider, which was cancelled in 1993), proposed accelerator projects in the 1950s and 1960s were enthusiastically supported. When the Stanford Linear Accelerator was completed in 1966, it was the largest single civilian scientific project ever constructed by the US government, yet it had earned wide support from both parties in both houses of Congress.[83] Fifteen years later, with Cold War pressures dissipating, the urgency of funding "Big Science" projects had waned.

Notes

1 Donald Hornig, "Conference on Research Administration in Colleges and Universities, American Council on Education," October 8, 1964, LBJ, DH, 8: Addresses and Remarks, 1964, p. 1.
2 PSAC, *Scientific Progress, the Universities, and the Federal Government*, GPO, 1960, pp. 7, 10.
3 An excellent example was the Society of Arcueil, a circle of French scientists who met regularly in the homes of Claude Louis Berthollet and Pierre Simon LaPlace a few miles outside of Paris in the first decades of the 19th century. See Maurice Crosland, *The Society of Arcueil: A View of French Science at the Time of Napoleon I*, Harvard University Press, 1967.
4 Abraham Flexner, *Medical Education in the United States and Canada: A Report to the Carnegie Foundation for the Advancement of Teaching*, Carnegie Foundation, 1910.
5 Gibbs to Slater, August 7, 1950, MIT, 155: NRC.
6 DuBridge to Flemming, August 12, 1953, MIT, 195: SAC.
7 Glenn Seaborg, "The University and Basic Science," *Chemistry and Industry*, March 2, 1957, p. 255.
8 Frederick Terman, "Fundamental Research in University and College Laboratories and Its Contribution to Industrial Research and Development," *Proceedings of the First Annual Northern California Research Conference*, January 12, 1949, pp. 34, 36.
9 PSAC, *Scientific Progress, the Universities, and the Federal Government*, p. 14.
10 Paul Samuelson to Rupert Maclaurin, 1949, IIIR, 43:5. Samuelson, one of the most celebrated economists of the 20th century, had been detailed to the Rad Lab during the war.
11 That said, Merle Tuve, the director of the Applied Physics Laboratory at Johns Hopkins, felt that running a large, grant-funded university lab was so time-consuming and administratively demanding as to preclude the possibility of doing highly creative work. In his words:

> There is a growing conviction among my friends in academic circles that the university is no place for a scholar in science today, because a professor's life nowadays is a rat race of busyness and activity, managing contracts and projects, guiding teams of assistants, and bossing crews of technicians, plus the distractions of numerous trips and committees for government agencies, necessary to keep the whole frenetic business from collapse.
> Quoted in Paul Forman, "Behind Quantum Electronics: National Security as Basis for Physical Research in the United States, 1940–1960," *Historical Studies in the Physical and Biological Sciences*, 18:1, 1987, pp. 196–97

12 Seaborg, "The University and Basic Science," p. 255.
13 See Waterman, "Research for National Defense," *Bulletin of the Atomic Scientists*, 9:2, 1953, pp. 37–38.
14 Richard Lewontin, "The Cold War and the Transformation of the Academy," in David Montgomery, ed., *The Cold War and the University*, The New Press, 1997, p. 29.

15 Rebecca Lowen, *Creating the Cold War University*, University of California Press, 1997, p. 68.
16 Donald Hornig to Conference on Research Administration in Colleges and Universities, American Council on Education, October 8, 1964, LBJ, DH, 8: Addresses and Remarks, 1964, p. 6.
17 Joseph W. Barker, "Grants," 1945, MIT, 179: Research Corporation, p. 2.
18 "Memorandum on Overhead Policy on Government Contracts," MIT, 164: ONR/OWA.
19 NSF, "Recommendations for a Uniform Policy for Paying Indirect Costs of Research Supported by the Federal Government at Universities and Colleges," June 1955, VB: 1913, pp. 1–3.
20 PSAC, "Scientific Progress, the Universities, and the Federal Government," November 15, 1960, GPO, Washington, DC, p. 30. Also, see Donald Hornig, statement at the Conference on Research Administration in Colleges and Universities, American Council on Education, Washington, DC, October 8, 1964.
21 Paul Doty, "Comments on Presentations Before the Panel on Basic Research," OSAST, 1: Basic Research, 12/57–11/60.
22 Frederick Terman, "Administrative Policies and Objectives of Research in Engineering Colleges," *Journal of Engineering Education*, 38:4, December 1947.
23 Donald Hornig, speech at the fifteenth anniversary of the NSF, May 27, 1965, DB, FA 965, 40: NSF, Loos Papers, 1951–1968.
24 Hornig, "Conference on Research Administration in Colleges and Universities, American Council on Education," October 8, 1964, p. 3.
25 DuBridge to Flemming, August 12, 1953, MIT, 195: SAC.
26 ODM-SAC, "The Role of the Scientists in Defense," May 30, 1951, DB, FA 965, 16: FCST.
27 SAC, "Revised Statement for Guidance," July 25, 1951, DB, FA 965, 16: FCST, p. 1.
28 Ibid.
29 Ibid., p. 5.
30 "Extract from a Report of the Basic Research Group, RDB," June 1953, MIT, 195: SAC, p. 2.
31 Bush to Baxter, Compton, et al., November 18, 1944, DB, FA 965, 38: OSRD, Bush, Van.
32 Quoted in, Alan Waterman, "The Role of the National Science Foundation," *AAAPSS*, 327, January 1960, p. 127.
33 Brookhaven was purposefully created in 1947 to give scientists in northeastern universities a regional lab in which they could do atomic work. While all of the national labs had expansive networks of university affiliates, only Brookhaven was truly operated by a consortium rather than one principal university. The consortium consisted of Columbia, Cornell, Harvard, Johns Hopkins, MIT, Pennsylvania, Princeton, Rochester, and Yale. From the start, Brookhaven was more committed to basic research than the other national labs, and by 1955, the AEC had declassified the Brookhaven reactor in an effort to make it more accessible to foreign scientists without clearance. For a journalistic account of its beginnings, see Benjamin Fine, "Laboratories for the Atomic Age," *NYT Magazine*, June 22, 1947. For more scholarly treatment, see Audra Wolfe, *Competing with the Soviets: Science, Technology, and the State in Cold War America*, Johns Hopkins University Press, 2013, particularly pp. 30–42.
34 From Peter Westwick, *The National Labs: Science in an American System, 1947–1974*, Harvard University Press, 2003, p. 2.
35 These figures are from an internal White House memo to the President. Buckley to Eisenhower, May 1, 1952, IIR, 43:3.
36 Henry Smyth, "Remarks to the General Information Meeting Banquet," Oak Ridge, October 25, 1949, DB, FA 965, 28: AEC 1949, p. 10.

37 Within the ORINS, the University of Tennessee was first among equals. One of the more important efforts of ORINS was in creating an AEC-sponsored hospital in which to investigate the health effects of radiation and the potential clinical uses of radioisotopes.

38 See Alvin Weinberg, "Some Problems in the Development of the National Laboratories," December 1955, EOL, 106963, p. 4.

39 Wolfe, *Competing with the Soviets*, p. 30.

40 Michael Aaron Dennis, "Our First Line of Defense: Two University Laboratories in the Postwar American State," *Isis*, 85, 1994, p. 431.

41 At least one critic of the arrangement questioned the need for university affiliation. Alvin Weinberg, the administrator of ORNL in the early 1950s, wrote to Glenn Seaborg,

> I am not thoroughly convinced that the big institutions in general really ought to be connected with the universities. The Lawrence Laboratory is an outstanding success, but it grew up very naturally and always was an organic part of the University of California. Do you think that the other big accelerator laboratories (such as Princeton-Penn) or the big Sherwood Laboratory at Princeton will be equally good examples?
>
> Weinberg to Seaborg, March 29, 1960, OSAST, 13: Basic Research

42 Roger Geiger, "Universities and National Defense, 1945–1970," *Osiris*, 7, 1992, p. 29.

43 Dennis, "Our First Line of Defense," p. 440.

44 "The Need for Higher Salaries for Research and Development Personnel," May 18, 1951, RG 255, NACA, IDC/4: General Correspondence, 4/7.

45 Interdepartmental Committee on Scientific Research and Development to the President, December 5, 1957, OSAST, 11: Interdepartmental Committee on Scientific Research.

46 Don Kash discusses the issue of Civil Service salaries in "Forces Affecting Science Policy," *Bulletin of Atomic Scientists*, April 1969, p. 12.

47 "Atomic Boss," *Time*, 53, January 3, 1949, p. 32, as quoted in Robert Seidel, "A Home for Big Science: The Atomic Energy Commission's Laboratory System," *Historical Studies in the Physical and Biological Sciences*, 16:1, 1986, p. 145.

48 Subcommittee on Scientific Personnel, Interdepartmental Committee on Scientific Research, "The Scientist in Government," RG 255, 1/IDC: Reports 2/5, p. 6.

49 See "Mobilization of Science with Regard Principally to the Physical Sciences in Universities," MIT, 194: SAC. Also W.H. Zinn, "Comment on National Laboratories," December 6, 1955, EOL, 106996.

50 Smyth, "The Role of the National Laboratories in Atomic Energy Development, *Bulletin of Atomic Scientists*, 6:1, 1940, p. 6, as quoted in Seidel, "A Home for Big Science," p. 148.

51 DuBridge to Riehlman, July 1, 1954, IIR, 43:3, p. 3.

52 Stuart Leslie, *The Cold War and American Science*, Columbia University Press, 1993, p. 14.

53 Quoted in Ibid., p. 14.

54 Slater to Conrad, October 24, 1945, MIT, 161: Nuclear Science and Engineering Laboratory.

55 Proposal, "LNSE at MIT," November 21, 1946, MIT, 161: Nuclear Science and Engineering Laboratory.

56 Minutes, November 13, 1946, "The Site for the 300 Mev Synchrotron," MIT, 161: Nuclear Science and Engineering Laboratory.

57 Bowen to Compton, August 30, 1946, MIT, 161: Nuclear Science and Engineering Laboratory.

58 MIT Press Release, January 19, 1950, MIT, 161: Nuclear Science and Engineering Laboratory.

59 "The Research Laboratory of Electronics at MIT," MIT, 181: Research Laboratory of Electronics.
60 Roger Geiger, "Universities and the National Defense, 1945–1970," *Osiris*, 7, 1992, pp. 32–33.
61 Kelly to Stratton, November 30, 1951, MIT, 181: RLE. Also, Daniel Kevles, "Cold War and Hot Physics: Science, Security, and the American State, 1945–56," *Historical Studies in the Physical and Biological Sciences*, 20:2, 1990, p. 251.
62 Paul Forman, "Behind Quantum Electronics: National Security as Basis for Physical Research in the United States," *Historical Studies in the Physical and Biological Sciences*, 18:1, 1987, p. 160.
63 Quoted in Leslie, *The Cold War and American Science*, p. 30.
64 Quoted in Ibid., p. 31.
65 See Michael Aaron Dennis, "Our First Line of Defense: Two University Laboratories in the Postwar American State," *Isis*, 84, 1994, pp. 427–55.
66 James Killian, *The Education of a College President*, MIT Press, 1985, p. 36.
67 Quoted in Leslie, *The Cold War and American Science*, p. 34.
68 Quoted in Ibid., p. 37.
69 Duncan to Killian, October 17, 1955, MIT, 156: Department of the Navy.
70 Reported in the *New York Herald Tribune*, May 28, 1951, as quoted in Forman, "Behind Quantum Electronics," p. 187.
71 My discussion on Stanford's postwar growth rests heavily on Stuart Leslie's work, *The Cold War and American Science*, Columbia University Press, 1993. To quote Leslie, "Or, as Terman liked to say, better one seven-foot high jumper on the team than lots of six-footers." (p. 45)
72 One measure of the extent of the success was that by the early 1950s Stanford attracted the second largest number (after MIT) of graduate students who had won NSF graduate fellowships.
73 Roger Geiger, "Universities and the National Defense, 1945–1970," *Osiris*, 7, 1992, pp. 33–34.
74 Kistiakowsky to McCone, February 5, 1960, OSAST, 11: High Energy Physics.
75 Many universities drew anti-war protesters during the latter years of the Vietnam conflict in the late 1960s, but this type of moral opprobrium was nearly unknown through the 1950s.
76 See PSAC and GAC, "An Explanatory Statement on Elementary Particle Physics and a Proposed Federal Program in Support of High Energy Accelerator Physics," OSAST, 11: High Energy Physics.
77 AEC Press Release, "Remarks by Glenn Seaborg at the dedication of the Princeton-Pennsylvania Accelerator," December 7, 1963, GS, 870:7.
78 The Madison reactor, which was to serve a consortium of regional universities known as the Midwest Universities Research Association (MURA), was never built. See Seitz to Fisk, January 27, 1958, OSAST, 11: High Energy Physics.
79 For budget figures, see National Policy Council on High Energy Accelerator Physics, "A Review of the Status of the High Energy Accelerator Field," September 15, 1958, OSAST, 11: High Energy Physics. Also, "A Ten-Year Preview of High Energy Physics in the United States," December 12, 1960, OSAST, 11: High Energy Physics.
80 National Policy Council on High Energy Accelerator Physics, "A Review of the Status of the High Energy Accelerator Field," September 15, 1958, OSAST, 11: High Energy Physics, p. 11.
81 Ibid.
82 AEC Press Release, "Remarks by Glenn Seaborg at the dedication of the Princeton-Pennsylvania Accelerator," December 7, 1963, GS, 870:7, p. 3.
83 "SLAC's Historic LINAC Turns 50 and Gets a Makeover," *SLAC News*, May 4, 2016, https://www6.slac.stanford.edu/news/2016-05-03-slac-historic-linac-turns-50-and-gets-makeover.aspx.

Figure 1 As head of the Office of Scientific Research and Development, Vannevar Bush
coordinated the government's research efforts during World War II. Later, he
conceived the postwar role for the government in supporting both basic and
applied research, which continues to guide the effort today.

Source: US Department of Energy

Figure 2 The Radiation Laboratory at MIT ("Rad Lab") was the site of radar development during the war. At its peak it employed over 4,000 scientists and engineers.

Source: MIT Museum

Figure 3 Lee DuBridge (center) directed the Rad Lab during the war. Later, he presided over the California Institute of Technology and served as science advisor to Presidents Truman, Eisenhower, and Nixon. He is shown flanked by visiting Welsh physicist E. G. Bowen (left) and staff physicist I. I. Rabi (right).

Source: MIT Museum

Figure 4 This remarkable photo of the gathered leaders of the National Defense Research Committee includes (from left to right) E. O. Lawrence, Arthur Compton, Vannevar Bush, James Bryant Conant, Karl Compton, and Alfred Loomis.

Source: US Department of Energy

Figure 5 Leslie Groves and J. Robert Oppenheimer at Los Alamos. The combination of Groves's administrative acumen and Oppenheimer's scientific leadership enabled the remarkable success of the Manhattan Project.

Source: US Department of Energy

Figure 6 The Manhattan Engineering District's efforts and resources were split between facilities at Oak Ridge, Tennessee (6), Los Alamos, New Mexico (7), and Hanford, Washington (8). Refinement of uranium and plutonium was done at Oak Ridge and Hanford; the bombs were assembled and tested at Los Alamos.

Source: US Department of Energy

Figure 7 Los Alamos, New Mexico.
Source: National Park Service

Figure 8 The complex at Hanford, Washington.
Source: US Department of Energy

Figure 9 Tennessee Valley Authority director David Lilienthal (9) and presidential advisor Bernard Baruch (10) recommended that all nuclear weapons be ceded to a new international energy organization after World War II. When that effort was blocked by the Soviet Union, Senator Brien McMahon of Connecticut (11) authored the bill which created the US Atomic Energy Commission.

Source: Library of Congress

Figure 10 Presidential advisor Bernard Baruch.

Source: Library of Congress

IN WASHINGTON IN 1949, Oppenheimer conferred with the late Senator Brien McMahon after defending Atomic Energy Commission against its critics.

Figure 11 Senator Brien McMahon of Connecticut.

Source: Oregon State Library Special Collections

Figure 12 Efforts to codify the recommendations contained in the Bush and Steelman reports led to a series of failed bills proposed by Senators Harley Kilgore of West Virginia and Warren Magnuson of Washington (shown here). Their efforts finally produced a workable proposal in 1949, which was signed into law the following year to create the National Science Foundation.

Source: University of Wisconsin Special Collections

Figure 13 The five-year delay in creating the National Science Foundation elevated the role of the Office of Naval Research in funding both basic and applied research in the immediate postwar years. Shown here is the most well-known building at ONR's main campus in Washington DC. By 1949, research funding from Department of Defense and the Atomic Energy Commission accounted for 96 percent of all federally funded research in the physical sciences.

Source: US Naval Research Laboratory

Figure 14 President Truman picked Alan Waterman, chief scientist of the Office of Naval Research, to be the founding director of the National Science Foundation. Waterman had earned the trust of the three constituencies most important to the success of the NSF: the executive branch, civilian scientists, and the military.

Source: National Science Foundation

Figure 15 Stanford's efforts to leverage federal research largesse to its own gain culminated in the construction of the Stanford Linear Accelerator (SLAC) in 1966. At least three different Nobel Prize winning discoveries can be traced to research performed at SLAC.

Source: US Department of Energy

Figure 16 The launching of Sputnik by the Soviet Union in 1957 opened the Space Age and impelled the United States to accelerate its own space program.

Source: NASA

Figure 17 The United States' initial foray into space through its Project Vanguard program initially produced a string of failures, including a launch pad explosion in 1957. Historian Enid Curtis Bok Schoettle recalled the nicknames of the failed Vanguard efforts: "Puffnik, Flopnik, Phutnik, Kaputnik, and Stayputnik."

Source: NASA

Figure 18 The US Army scored a coup in importing German engineer Wernher von Braun to head its rocket program. Building off of his work on the Nazi's V-2 rockets, von Braun helped the army to develop its Redstone rocket program.

Source: NASA

Figure 19 The United States would ultimately celebrate a series of accomplishments in space, such as John Glenn's circumnavigation of the Earth in 1962, Ed White's space "walk" outside of his Gemini capsule in 1965 (shown here), and Neil Armstrong's moon landing in 1969.

Source: NASA

7 Science, the Hot War, and the Race to the Hydrogen Bomb

Keeping Up With the Soviets

On August 29, 1949, the Soviet Union exploded an atomic bomb at the Semipalatansk test site in Kazakhstan. The incident intensified American concerns over Soviet scientific and technological supremacy and accelerated efforts by the United States to strengthen scientific research and training.[1] The fear was threefold: that the Soviet Union was surpassing the United States in basic scientific research and capability; that the Soviets were surpassing the United States in industrial production and military strength; and that the Soviets would soon supersede the United States as the world's preeminent nuclear power. Coupled with a broader fear of Soviet expansionism, American military and scientific planners leveraged the event to argue for increased efforts in scientific recruiting and training and more funding for basic and applied research (particularly in the military sphere). Many political and military leaders also responded by lobbying for production of greater amounts of fissile material, greater numbers of atomic bombs, more efficient and destructive atomic bombs, and development of a new fusion-based bomb variably described as a "super", an H-bomb (hydrogen), or a thermonuclear device. A 1950 report from the AEC to President Truman described the Soviet Union as, "animated by a new fanatic faith, antithetical to our own, and seek[ing] to impose its absolute authority over the rest of the world."[2] The response to such a threat lay in nurturing and developing scientific talent, developing superior weaponry, and accelerating weapons production.

For all of the criticism of the Soviet system — the mediocrity of consumer goods, the suppression of free thought and creative endeavors, and the outright banishment of political dissent — American analysts recognized the ability of the Soviet state to mobilize quickly and forcefully to thwart external threats and to direct resources and personnel where national need might dictate. Although the United States had more trained scientists (in the physical sciences) in 1950, the Soviets were rapidly closing the gap. Since the Bolshevik Revolution of 1917, the Soviet Union had increased the number of universities tenfold and the population of college and university students fifteenfold. (The United States had doubled its university count

DOI: 10.4324/9781003363897-7

during this time while increasing its university student population fivefold.) Forty-five percent of university students in the USSR studied science versus 30 percent in the United States, and the country was producing 40,000 new engineers annually by 1954 (versus 23,000 in the United States). For those students who went on to earn a doctoral degree, nearly 80 percent of the Soviet students earned their degrees in science or engineering, versus only 50 percent of American students.[3] Alan Waterman, the director of the NSF, wrote that year, "Thus, the indications are that we are already unable to match Russia in our output of engineers."[4] Howard Bevis, the chairman of the President's Committee on Scientists and Engineers, concurred, warning, "The rate of Russian progress in most scientific fields is so rapid that, unless we broaden and strengthen our own efforts, there is little question of Soviet superiority five or ten years from now."[5]

What of quality? While it was very difficult for American observers to discern the quality of the Soviet education system, and while the biomedical sciences seemed prone to manipulation by communist ideology, Russians' accomplishments in the applied physical sciences suggested a high level of facility and mastery. In the first decade after World War II, the Soviets had exploded several nuclear bombs, built enormous infrastructure projects (such as the Gorki Dam and the Volga-Don Canal), and mastered jet propulsion. The Soviet Mig-15 jet fighter was superior to anything that the Americans possessed. The Soviets were reverse engineering agricultural and industrial equipment, mastering digital computing, and raising the quality of their metal-working. Soviet solid state physics, electronics, low-temperature physics, nuclear physics, and aeronautical engineering were all excellent.[6]

The Soviets were piggybacking their scientific prowess on a highly targeted national system of public education designed to serve the interests of the state. In contrast to American education, which tended to cater to the preferences of the individual student, the Soviet system identified gifted students in specific disciplines deemed of interest to the state at a young age and pushed them into paths of excellence. Observers of Soviet education noted its "aggressive" and "tough-minded" qualities; it "ruthlessly weeded out the mediocre students" in an effort to populate universities with the best of a generation. James Killian, the President of MIT, described Soviet scientific education as part of a broader cultural posture, "that education is essential to the progress of the Soviet people, a 'mighty weapon in the cause of Communism.'" This stress on education, Killian continued, was "the most important secret weapon, and the success they are having in promoting education may be one of the greatest of their threats to the free world."[7]

American Sovietologists feared the Soviet zeal. Envisioning a nation of ideologically committed Communists, a group of scientists and university leaders on an NSF-sponsored tour of the USSR saw "determination and a national spirit on the part of the people which seems to be relatively absent from the American scene." The Russian people aspired to "world leadership" in science and technology, of achieving "world supremacy without the need of military domination."[8] Lloyd Berkner, the president of the Associated Universities and a consultant on scientific affairs to the Department of State, acknowledged the great strides the Soviets had made by the 1950s, including building the largest proton accelerator, the most intense neutron source, the largest optical and radio telescopes, and a plethora of medium- and long-range rockets. The country was planning a comprehensive space program, with the long-term goal of lunar and planetary landings. Berkner concluded, "The Soviet program is so bold as to invite frequent failure, but it has the stuff from which achievement rises. Scientific leadership in the West must be equally bold and imaginative if it is to compete successfully."[9]

In assessing the risk of Soviet scientific supremacy, America's scientists led its politicians. While many American physicists correctly discerned the nascent power of Soviet mathematics, chemistry, and theoretical physics during the late 1940s, members of Congress tended to brush warnings off as alarmist. The NSF budget continued to lag that of the defense services, and Congress had yet to seriously engage with broad shortfalls in American secondary education. Even as more Americans attended college courtesy of the G. I. bill, few American school districts devoted substantial resources to serving the nation's most gifted and promising students or singling out those students for supplementary opportunities in science and engineering. One of the few who raised the alarm in the 1950s, Representative Sterling Cole from New York, lectured his colleagues:

> Yet the facts are these: The Soviets have assigned top priority to building up a vast reservoir of scientific manpower. This year, the Soviet educational system will graduate more scientists and engineers than will our own universities. Projections for the next several years indicate that the gap between the number of scientific and technical graduates in each of the two nations will widen – in favor of the Soviets. By 1965 – if present trends continue unchanged – American science may be forced to yield world leadership to the Soviets. Since military supremacy is today almost synonymous with scientific supremacy, the balance of world power may thereby shift decisively in favor of the Soviet Union.[10]

Parallel to their push in science and engineering, the Soviets expanded the size and scope of their military. While it was true that in 1950 the United States outproduced the Soviet Union fourfold along many axes of industrial production – steel, aluminum, electric power generation, crude oil – the Soviet Army and Air Force were larger than their American counterparts and growing increasingly sophisticated. By the mid-1950s, the Soviet Air Force was capable of long-distance bombing runs; the Soviet Navy's submarine fleet patrolled all of the world's oceans; and the nation's guided missiles could hit NATO targets.[11] Investing triple the portion of GDP in its military relative to the United States, the Soviet Union was developing a military sized more appropriately for offensive actions in western Europe or in the Middle East rather than for border defense. One chilling report to the White House warned that Soviet military growth was consistent with a "design for world domination."[12] Senator Henry Cabot Lodge (R., MA) warned President Eisenhower in 1951,

> [T]he cold, brutal fact is, Mr. President, that the United States does not have air supremacy, air superiority or anything like it. The staggering fact is that on balance, air superiority as well as land superiority lies with the Soviet Union.[13]

Lodge recommended that the United States immediately add 3,600 planes to its own Air Force and Navy and supply 6,000 military aircraft to its NATO allies.

While the Soviets lacked the outright industrial capacity to overwhelm the United States in an all-out war, its system of moving basic research to military application seemed more efficient. James Killian, a member of the SAC, warned of the Soviet's accelerated time line in producing operational military hardware – getting new military aircraft designs to production three years faster than Americans could, for example.[14] To American analysts, the Soviet Army appeared to make decisions and strategic plans in closer cooperation with the nation's scientific leadership, placing scientists in senior military advisory posts to help generals shape military planning around emerging technology. Killian would write later that decade, "One of the most urgent needs in our whole defense organization is for men, whether they be in uniform or not, who understand the integration of systems and the organizational implications inherent in our new weapons technology."[15]

To some degree, these stark estimates overstated Soviet capacity and capability at the time. Soviet factories had little excess capacity, while production of military hardware in the United States could be quickly expanded if necessary. Many of the Soviet Union's planes, ships, and submarines were not as technologically sophisticated or robust as their American counterparts. The Soviet Union's thousands of engineers tended to lack deep grounding in the basic sciences – more akin to an American technician than an engineer. American engineers tended to have broader academic training and deeper scientific grounding, making them more flexible in approaching engineering challenges.[16] Leon Trilling, a professor of aeronautical engineering at MIT who had studied the Soviet system, viewed Soviet engineers as narrow and brittle. He wrote of them, "The full cross-section of engineers in the Soviet Union have apparently not yet acquired that degree of engineering 'feeling' which only broad familiarity with machinery can bring. They work by the book and require detailed direction."[17]

In some areas, the United States far outpaced the Soviets. While the Soviet Union probably possessed no more than 200 digital computers by the late 1950s, the United States had more than 3,000. American transistor

technology led Soviet technology substantially.[18] The United States produced better guidance systems and radar technology. Its planes flew farther
and its ships sailed faster. Still, the Soviet were constantly improving. Vannevar Bush, who had become somewhat of a shrill alarmist by the mid-1950s,
watched Soviet progress closely and concluded that while the United States
still led technologically, it could not reduce investment in military research
and development. In a general report from that time, Bush wrote,

> We need to be in the lead both in means of stopping enemy bombers
> and in means of penetrating enemy defense. There is a decided possibil-
> ity that the defense may catch up, that radar warning nets, antiaircraft
> artillery, ground-controlled interception utilizing jet-propelled aircraft,
> and ground-to-air missiles may make it increasingly impractical to pen-
> etrate to prime targets.[19]

In sum, while the Soviets continued to dominate in the sheer size and scale
of their Army and Air Force, the United States probably maintained a technological edge through the 1950s, albeit a shrinking one. The Soviets' ability
to divert both personnel and resources to specific national priorities (such as
military supremacy) eased the path from research to implementation. They
were capable of excellent work in the mathematical and physical sciences,
and they had rapidly expanded the pipeline of talented young people entering science and engineering programs. And unlike the United States, which
faced persistent populist pressure to rein in military spending and decrease
the size of the standing Army, the politburo could operate with a degree of
political impunity.[20]

But most chilling was growing Soviet atomic capability. The Soviets' successful 1949 atomic bomb detonation was only the first visible sign of a
massive investment in nuclear technology. Soviets were building up gaseous
diffusion technology to refine uranium.[21] Estimates by the Central Intelligence Agency in 1950 suggested that the Soviets already possessed an armory
of 10–20 atomic bombs, which would grow to 200 by 1954.[22] The Soviet
Union already possessed a long-range bomber capable of delivering the
bomb to eastern American cities. Representative Henry ("Scoop") Jackson
(D., WA), a hawkish member of the Joint Committee on Atomic Energy,
warned his colleagues that only atomic superiority had held the Soviets in
check since 1945, and that now that the monopoly was broken, Americans
must commit themselves to producing and maintaining the world's foremost
nuclear arsenal including smaller, tactical warheads, which could be used to
surgically neutralize enemy armies. "I believe that we must immediately and
dramatically expand the scale and scope of our atomic effort," Jackson lectured his congressional colleagues. "In place of approximately $1 billion we
are spending on atomic weapons this year, I propose that we now undertake
to spend between 6 and 10 billions [sic] annually on this supreme deterrent
against Kremlin aggression."[23]

In response, the United States sharply increased spending on nuclear arms development. In 1952, the AEC announced a 50 percent increase in plutonium production with the goal of producing an array of fission bombs of various sizes which could be used in a tactical rather than strategic fashion – that is, for hitting specific targets – with the idea that the Army would also integrate nuclear weapons into its ordnance. Secretary of Defense Robert Lovett suggested that increasing the budget for nuclear arms from the current $1 billion to $6 billion would allow the combined armed services to outproduce the Soviets and maintain a comfortable atomic lead.[24] The increase was impelled by a very real fear of evolving Soviet supremacy – "Better to scrape the bottom of the tax barrel than to scrape atomic rubble from the streets of New York," warned Senator McMahon – but also by the strategic recognition that atomic weapons simply made sense amid the current doomsday scenarios. Atomic bombs were terrifying and awesome and scrambled conventional scenarios of total war. Atomic capability meant not having to match the enemy in troop strength. The weapons were inexpensive, too, relative to the total cost of the Cold War. By 1952, total spending on nuclear arms constituted only 3 percent of the US defense budget.[25]

By 1952, the United States was producing 50 bombs per year, and the AEC was expanding plutonium production and gaseous diffusion capacity *en route* to becoming the nation's single largest consumer of electric power. At Los Alamos and the Berkeley Rad Lab, AEC-funded scientists worked to improve the efficiency of the devices. While only about 1 percent of all of the fissionable material in the two bombs dropped on Japan had actually undergone fission, a newer bomb detonated on the Pacific island of Eniwetok in 1949 had probably achieved closer to a 50 percent fission rate. At the same time, the United States was building a newer, more efficient breeder reactor in Idaho to produce plutonium more reliably and at a lower price.[26] The effort seemed reasonable and proportional. A White House advisory group recommended to President Truman that the only rational response to a Soviet nuclear stockpile was a countervailing and superior American one – an atomic posture which would "act as a deterrent to war."[27]

Ethical Questions and Testing

Amassing an atomic arsenal raised ethical questions for the nation and forced an examination of the use of atomic weapons on Japan. Despite the devastation and death that had resulted from the blasts, many of the men who had been deeply invested in the development of the atomic bomb, including all who sat on the interim Stimson Committee, which advised Truman on using the bomb, felt that its use had been justified. The Stimson Committee believed at the time that using the bomb could possibly prevent 2,500,000 military fatalities – the frequent estimate for total soldiers lost in the event of an invasion of Japan (including Japanese combatants).[28] Death from an atomic strike was not qualitatively worse than death from conventional bombs and bullets. Any eventual victory would have rested on destroying industrial centers responsible for producing arms and munitions, and such centers were embedded within Japanese population centers. Writing in September 1945, James Bryant Conant, a member of the Stimson Committee, asserted that he saw little moral difference between victory through use of innovative weaponry and the traditional approach of "mangling and maiming men."[29] Similarly, Arthur Compton, chancellor of Washington University (and brother of Karl Compton), explained,

> We made the best choice for man's future that we knew how to make. . . . The chief hope in the atomic bomb was that by a dramatic demonstration of its power the enemy would become convinced of the uselessness of further fighting.[30]

Moreover, the second bomb (on Nagasaki) had been justified by the need to demonstrate that the United States could sustain its atomic attacks. Given that the axis powers (particularly the Germans) were knowledgeable about the difficulties of enriching a uranium core, it had been critical to demonstrate that the United States had the technological ability to continue to produce atomic bombs at a rate which would be decisive to winning the war. Compton explained the following year that

> the building of bombs would be so slow as to be of little military significance was in fact the conclusion to which the German scientists had come. It was necessary for us accordingly to give the impression of having available an indefinite supply.[31]

But even those who had advocated dropping the bombs on Japan were alarmed at the accelerating arms race. Hiroshima and Nagasaki had rapidly become emblematic of humans' capacity for destruction, and a steady buildup of the nation's atomic arsenal seemed not so much a rational strategy as a step toward madness. The United States initiated research programs on the long-term biological effects of the atomic bomb blasts soon after the Japanese surrender and discovered, to no surprise, that the large radiation release caused serious injury and death months later. While the bombs had killed an estimated 125,000 people at the moment of detonation, physicians estimated the true death tally at closer to 200,000. The blasts had wiped out economic activity, undermined ordinary life for the survivors, and created a stream of atomic refugees. John Bugher, the deputy director of the AEC's Division of Biology and Medicine, noted in his report, "In contrast to the heavy bombing of cities, such as Tokyo, which ultimately produced nearly equivalent destruction, the cities of Hiroshima and Nagasaki ceased to exist as functioning entities."[32]

American military physicians conducted multiyear studies on Japanese survivors proximate to the blasts and noted increased rates of both miscarriage and abortion. Prevalence of leukemia was substantially elevated among people who had been within 2,000 meters of the bomb blasts, as were the frequency of cataracts and pediatric dental malformation. White blood cell counts in nearly all who had been near the centers of the blasts were substantially depleted, although they recovered over time. Of particular interest were genetic abnormalities, which could only be observed as the next generation aged. Exposure to high levels of gamma and neutron radiation in laboratory settings elevated genetic mutation in plants, but it was impossible to infer from these observations effects on humans. Physicians observed a pool of 5,000 babies born in the immediate aftermath of the war and recorded both genetic changes and phylogenetic manifestations, but true damage would not be known for some time.[33]

In 1947, the victorious powers established the Atomic Bomb Casualty Commission to continue the medical studies of both survivors and their offspring. Within a few years, researchers were able to conclude that the injured had recovered from some initial radiation effects: blood counts had returned to normal, hair had grown back, and healthy sperm production had rebounded. The rise in rates of cataracts had continued unabated, although most of the cataracts remained very small. But long-term effects on survivors eluded researchers, who confronted both natural mortality (it was often impossible to determine if a bomb survivor later died from radiation or another cause) and high rates of internal migration. By 1950, of the 80,000 surviving Hiroshima residents and 15,000 surviving Nagasaki residents who had lived within 2,100 meters of the bomb blasts, only about 28,000 remained; the rest had all moved on.[34]

Even if ethical questions could be dismissed, logistical questions remained. Atomic bombs were relatively cheap to build but quite expensive to deliver. They required the services of long-range bombers which were vulnerable to radar, antiaircraft guns, and increasingly accurate guided missiles. The allied bombing strategy which had proven so effective against Germany during World War II was probably obsolete as early as 1950. Bush, very much the realist, wrote in 1949 that, "the specter of great fleets of bombers, substantially immune to methods of defense, destroying great cities at will be atomic bombs is a specter only," and that the bombing runs conducted so effectively over northern Europe were now "obsolete against a fully prepared and alert enemy."[35] Radar continued to improve, while evasive measures had hardly developed. Bush explained:

> It is now fairly well accepted that the lone bomber setting out to fly an atomic bomb at high altitude many hundreds of miles over the territory of a fully alert and equipped enemy would never get to its target. The movement of every plane in the country would be continuously charted and followed by radar. Against an unknown plane, pursuit planes would take off in haste and numbers. They could without much doubt bring it down. Perhaps it could sneak in at low altitude, but that would also be risky and difficult. Accompaniment by fighters would not help much.[36]

Through the 1950s, both Presidents Truman and Eisenhower wrestled to avoid an arms race and nuclear Armageddon. As early as 1951, scientists associated with the Federation of American Scientists called on Truman to work toward international control of atomic arms by calling on "imaginatively and patriotically motivated men."[37] In his farewell address, President Eisenhower warned sharply of the developing "military industrial complex" whose enormous influence could "endanger our liberties or democratic processes."[38] In a private follow-up meeting with his science advisor George Kistiakowsky, Eisenhower explained his concern that the arms race would undermine the motives and integrity of the nation's scientific research and that the military's research budget had created "a most dangerous combination of special interests which is of real danger to the future of our free society." Moreover, Eisenhower feared the corruption of elite educational institutions, committed as they were to free inquiry and new scientific knowledge, which now competed for military research contracts and were "influencing research people on their staff to abandon basic science for the sake of higher monetary rewards."[39]

Even Premier Nikita Khrushchev of the USSR expressed concern for the growing arms race which threatened to hamper his nation's economic growth. In 1958, the premier wrote directly to Eisenhower to request that the United States cease high-altitude bomber flights over western Europe and the Artic and create a wide "no-fly" zone which could possibly ease tensions and to show "practical initiative" in ameliorating growing danger of nuclear conflict.[40] Such efforts, however, were not the norm. Formal arms limitation talks (SALT I and II) would not begin until 1969, and the formal Strategic Arms Reduction Treaties (START) would not be signed until 1991 and 1993.

A secondary strategy in reducing harm caused by atomic weapons lay not in limiting the production of the weapons but rather in limiting their testing. Atomic blasts released substantial amounts of radiation while also widely scattering radioactive debris. Concern over the health effects of testing to "downwinders" – people living in the path of post-test radioactive fallout – grew after the first of the hydrogen bomb tests in 1954. The largest of the experimental hydrogen bombs, code named BRAVO, scattered radioactive debris over a nearby Japanese fishing boat in the Pacific Ocean, afflicting 23 crew members with severe radiation illness and ultimately killing one.[41] In 1957, the Soviets proposed a three-year testing moratorium, but the proposal was rejected by Eisenhower. Only in 1958, when the AEC's General Advisory Committee determined that the tests were posing a "crisis", did the United States concede that its nuclear arsenal was sufficiently robust as to obviate the need for additional testing.[42] That year, the two nations sent representatives to meet in Geneva to discuss a comprehensive test ban. The proposed agreement – a ban on atmospheric tests but the continued use of underground tests – limited environmental exposure to radiation poisoning while doing little to ease tensions between the two nuclear superpowers.

Underground tests were very difficult to detect through the use of seismological equipment, and spy plane photos were of little use. Negotiations largely ceased after CIA pilot Francis Gary Powers was shot down over Sverdlovsk in May, while the American elections that Fall effectively put off further negotiations.[43] Upon taking office, President Kennedy refrained from holding further atmospheric tests but threatened to resume them should the Soviets test first.[44]

Atomic arms disturbed the sensibilities of many but were deemed by military leadership in both the United States and the Soviet Union necessary in responding to military threats. Although a few scientists, notably Oppenheimer, expressed concerns over the moral rectitude of developing and testing such weapons, political leaders in both the United States and the Soviet Union committed themselves through the 1950s to amassing more powerful and more efficient nuclear weapons and to developing more reliable delivery vehicles to ensure that the weapons could reach their targets. In the two decades after the end of World War II, a nuclear arms race appeared to be a reasonable response to rising existential threats. The response would continue to power much investment in both the physical and the engineering sciences into the following decade and in fact would redefine some of the more compelling foundational questions in physics.

The "Super"

The obvious response to the escalating arms race was a newer type of weapon based not on splitting heavy atoms but rather on fusing light ones. Fusion reactions powered the sun and stars and, if done in a controlled way, could potentially produce bombs 1,000 times as powerful as the bombs which had been dropped on Japan. The promise of the new technology was twofold: (1) the material powering such a weapon would be hydrogen, the most abundant element in the universe, and (2) unlike in fission bombs, where the initial fission reactions tended to dissipate the remaining uranium or plutonium before it had a chance to undergo fission itself, a fusion bomb could potentially be made infinitely large. There were several substantial barriers to building such a weapon, however. Ordinary hydrogen, which was simply a single proton orbited by an electron, underwent fusion reluctantly. It fused far more easily when it was enriched with deuterium, a form of hydrogen in which each atom contained an extra neutron, or even tritium, in which each atom contained two extra neutrons. Hydrogen found in nature contained a fair amount of deuterium (about 7.5%), but tritium actually needed to be synthetically produced by reacting hydrogen with lithium. The second barrier to producing a fusion reaction was in actually initiating the reaction, which happened only under conditions of enormous pressure and at very high temperatures – conditions found at the center of a star. Such conditions had been previously impossible to obtain. Once the fission bomb came into existence, a fusion bomb became theoretically possible.

In the aftermath of the Soviet Union's detonation of an atomic bomb in 1949, the General Advisory Committee of the AEC recommended to President Truman that the United States begin developing a fusion weapon, which became known as either the "Super" or hydrogen bomb. The decision to build such a bomb raised the usual logistical questions surrounding cost and schedule but also raised troubling questions about the ethical ramifications of such a weapon. While the atomic bombs which destroyed Hiroshima and Nagasaki had leveled approximately two square miles of those cities' cores, a hydrogen bomb could potentially level an area of 120 square miles — that is, an entire major city. The difference, illustrated graphically in various mock-ups, was between a bomb which could level about 30 square blocks of midtown Manhattan versus a bomb which could level the entire island of Manhattan (13 miles long) as well as most of Queens and the Bronx. This was destruction on a scale hitherto unknown — a weapon which could conceivably kill ten million people at one time or 100 times as many as the combined toll of the two bombs dropped on Japan. Oppenheimer warned of the moral cost of building such weapons: "In some sort of crude sense which no vulgarity, no humor, no overstatement can quite extinguish, the physicists have known sin; and this is a knowledge which they cannot lose," he wrote.[45]

Members of the Truman administration debated moving forward with the Super. Arguing against the bomb were a varied group of physicists, engineers, military analysts, and politicians. Hans Bethe, the Cornell University physicist who had directed the theoretical division at Los Alamos within the MED, compared use of such a weapon to the genocidal warfare of Genghis Khan. "If we have learned any lesson from the aftermath of World War II," he wrote, "it is that physical destruction brings moral destruction."[46] Oppenheimer questioned the tactical value of a weapon which could really only be optimally used on a handful of very large cities and then only for near-genocidal purposes. Its value seemed to be more psychological than strategic, answering a deep need to trump Soviet aggression once and for all. He wrote to James Bryant Conant, the president of Harvard,

> What does worry me is that this thing appears to have caught the imagination, both of the congressional and of the military people, as the answer to the problem posed by the Russian advance . . . that we become committed to it as the way to save the country and the peace appears to me full of dangers.[47]

In support of the bomb's development were two powerful postulates: (1) that *somebody* was going to develop the weapon, and it would be far better if the United States developed it first; and (2) that the true enormity of the weapon might actually be a strike for peace, as no country could rationally initiate a nuclear attack under threat of a hydrogen bomb counterstrike. These two arguments tended to persuade senior members of the AEC's

General Advisory Committee, and more hawkish members of Congress, to move forward. Ernest Lawrence, the *pater familias* of the nuclear physics community, warned David Lilienthal, the chairman of the AEC, in 1950, "If we don't get this super first, we are sunk. The U.S. would surrender without a struggle."[48] President Truman's own security advisors concurred, informing him in a White House report in 1950 that beating the Soviets to the weapon would allow the United States to dictate the rules of engagement.[49] And Brien McMahon, the most vocal of the anti-Soviet senators, queried the President in a lengthy letter,

> What happens if supers are aimed at New York, Chicago, Los Angeles, and Washington? Will we possess our own supers, ready to retaliate in kind and to throttle the attack at its source – or will we lack such weapons and suffer defeat and perhaps utter annihilation as the result?[50]

On November 25, 1949, the AEC recommended to Truman that the effort to build a hydrogen bomb be allowed to proceed. Lewis Strauss, the commission's chairman, prefaced the recommendation with his own preference:

> I believe that the United States must be as completely armed as any possible enemy. From this, it follows that I believe it unwise to renounce unilaterally any weapon which an enemy can reasonably be expected to possess. I recommend that the President direct the Atomic Energy Commission to proceed with the development of the thermo-nuclear bomb, at highest priority.[51]

The AEC's recommendation was grounded purely in strategic parity: the weapon's feasibility (better than 50%), knowledge that the Russian's were working on their own Super, the pressing timeline (Russians might have the bomb within fewer than five years), and the potential for such a large weapon to truncate the growing arms race.[52] Admiral Sidney Souers of the National Security Committee summarized his support more starkly. "It's either we make it or we wait until the Russians drop one on us without warning."[53]

But countering these recommendations was a dissent penned by a small subgroup of the General Advisory Committee. The authors of the minority report, which included Oppenheimer, Conant, DuBridge, and Oliver Buckley, wrote that the dangers of the Super outweighed any tactical advantages that it might confer. "Let it be clearly realized that this is a super weapon," they wrote.

> The reason for developing such super bombs would be to have the capacity to devastate a vast area with a single bomb. Its use would involve a decision to slaughter a vast number of civilians. . . . Therefore, a super bomb might become a weapon of genocide.[54]

Two others in the minority, Isidor Rabi and Enrico Fermi, reiterated the dissent in their own letter: "By its very nature, it cannot be confined to a military objective but becomes a weapon whose practical effect is almost one of genocide."[55] Truman, however, sided with the majority, and on January 31, 1950, directed the AEC to proceed with the development of the hydrogen bomb.

Work progressed quickly. By the following year, the Berkeley and Los Alamos labs were able to produce a first-generation fusion device ("George"), which generated a small fusion blast even while drawing the bulk of its destructive power from a fission bomb trigger. The following year, the United States exploded its first true hydrogen bomb, a ten-megaton device (equal to 10 million tons of TNT) in the Eniwetok Atoll in the south Pacific. The prediction of a 1,000-fold increase in destructive power had been born out fairly accurately, as the original Hiroshima bomb had only achieved the destructive power of 13 kilotons of TNT. This test was followed by the rapid detonation of six hydrogen bombs in 1954, constituting a program titled Operation CASTLE. The first of these six, the Bravo test, was a device of nearly 15 megatons – the largest device that the United States would ever detonate. It was followed by Romeo, Koon, Union, Yankee, and Nectar, varying between 100 kilotons and 13.5 megatons. The following year, on November 23, 1955, the Soviet Union exploded its own thermonuclear bomb. Chairman Nikita Khrushchev, on an official state visit to India at the time, boasted, "I shall not say there has not been such an explosion. It was a terrific explosion."[56] Soviet officials claimed it to be the largest nuclear device ever detonated, but American analysts were skeptical. Herbert York, a veteran of the Manhattan Project and the founding director of the Lawrence Livermore lab, wrote that while he did not know the precise power of the first Soviet hydrogen bomb, he estimated it to have one-third the power of Mike, the first American thermonuclear device from 1952.[57]

The Super project had not challenged American physicists at the level that the Manhattan project had. Once American scientists were in control of fission, the principle of initiating fusion was fairly clear. The most challenging component of the project, creating enough tritium to catalyze the fusion reaction, had been accomplished fairly quickly with adequate supplies of lithium. The entire effort had cost no more than $300 million and had taken only two years from the time that the President gave his approval.

Work on the hydrogen bomb had divided the physics community. Oppenheimer was stripped of his security clearance in 1954, officially for youthful dalliances with communism but at least partially for his opposition to the Super. The physicist most identified with the bomb, Edward Teller, would be ostracized by the physics community in later years for his testimony against Oppenheimer and for his unapologetic advocacy for the Super. Shortly after the first successful tests, public support for nuclear weapons began to wane; the destructive power of such weapons simply overwhelmed tactical logic or moral comprehension. Even Vannevar Bush, almost fanatically anti-Soviet and militaristic in his later years, admitted that while he and his allies had argued vigorously for the development of the Super, "we nevertheless searched assiduously for a way out of the morass."[58] He concluded, in a 1954 letter to Conant,

> There is some hope in my mind that after we have lived with this sort of thing in the world for a while, the general attitude everywhere may change and agreements of some sort may indeed by feasible. In fact, if this does not occur we face a very appalling future indeed.[59]

Birth of PSAC

In 1951, President Truman established the Science Advisory Committee (SAC) and appointed Oliver Buckley, the former president of Bell Labs, chairman. The committee, with part-time members and no permanent staff, served to keep the President advised on scientific issues relevant to his leadership and to coordinate research efforts across the federal agencies. It reported to the President through the Office of Defense Mobilization and thus tended to have a less potent effect than was originally planned, particularly when competing against the influence of the autocratic Secretary of Defense, Charles Wilson.[60]

Although the committee boasted notable members – Conant, Killian, Oppenheimer, and Waterman among them – it lacked the influence of a full-time representative within the permanent White House staff.[61] Over the next half decade, the SAC deliberated about structural changes which could allow the President to be better informed about scientific matters while also forcing the various federal agencies involved in research to coordinate their work so as to avoid redundancy. The committee floated the idea of an executive-level Department of Scientific Research but dismissed it as politically unfeasible given its clear incursion into DoD turf. Looking to the wartime success of the OSRD in fulfilling that role, the committee concluded that that agency's success had resulted from the concurrence of wartime exigency and the singularly powerful figure of Vannevar Bush. Moreover, Bush had been fortified with a large permanent staff with which to carry on his work. "The respect which the recommendations of these leaders received was, without doubt, due more to their personal abilities and standing than to any authority with which they were clothed," the SAC concluded.[62]

Although the SAC was able to produce two fairly significant documents in its short life – the Technological Capabilities (Killian) Panel report and the Security Resources (Gaither) Panel report – it lacked the political suasion to truly coordinate between competing agencies.[63] Through the 1950s, the White House consistently underestimated growing Soviet nuclear capability and failed to adequately coordinate weapons development work between agencies. Notably, at one time all three uniformed services were developing their own rocket systems leading to a divided satellite launch effort at the end of the decade. (The nation launched three different satellites on two different rocket systems over five months in late 1957 and early 1958.)[64] In describing the effect of this loose system of coordination on the development of military hardware, Rabi responded succinctly, "Well, what happened was that we got licked."[65]

In response to American panic after the launch of Sputnik in 1957 (see Chapter 9), Eisenhower reconfigured the SAC into the President's Science Advisory Committee (PSAC), appointed James Killian of MIT chairman, and situated the apparatus directly in the White House rather than in the ODM. At the same time, he established a new Office of the Special Assistant to the President for Science and Technology (OSAST) and appointed Killian to this position as well. The dual appointment served to link the panel of distinguished scientists more directly to the President in an effort to embed a scientific dimension within all executive decision-making.[66] Notably, PSAC and OSAST shared a staff, and Killian's role as the first science advisor inspired confidence throughout the scientific community. In recruiting Killian to the position, Lee DuBridge, chair of the SAC, told him, "On this there has been no argument because there is no second choice . . . namely, that you yourself be asked to serve."[67]

From the beginning, PSAC devoted more than half of its time to security issues. Its Arms Limitations and Control Panel, Limited Warfare Panel, and Space Science Panel all sought to inform strategic planning with the latest scientific knowledge and insight. It addressed such topics as space exploration, missile and satellite development, and nuclear weapons technology, all while vigilantly observing scientific and technological progress within the Soviet Union. It created standing committees to study the science and technology of ballistic missiles (chaired by Jerome Wiesner), anti-submarine warfare (Harvey Brooks), arms limitation (Killian), graduate education (Glenn Seaborg), air defense (Emmanuel Piore), high-energy physics (also Piore), and space science (Edward Purcell). Overall, it strived to coordinate development work across the military branches, articulating broad strategic military objectives rather than fractured goals for the individual services, and pushing the armed services to rethink their fundamental approach to warfare.[68] In so doing, it took seriously the conclusion of the 1953 Hoover Commission Report on Research and Development, which stated that the armed services "have not distinguished themselves in the initiation of radically new approaches to weapons systems."[69]

PSAC consistently recommended increases to the nation's basic research budget – a recommendation which had stood consistently since Vannevar Bush had so recommended in his 1945 report, *Science, The Endless Frontier*. A 1960 report by Glenn Seaborg argued that basic research was of "absolutely critical important to the national warfare."[70] But, as historian Audra Wolfe writes, what changed in 1957 was "the willingness of policymakers to take their advice."[71] Within a year of the creation of PSAC, the NSF budget had tripled, and by 1960, it had doubled again. At the same time, the concentration of funding within the nation's elite research universities grew more pronounced under the guidance of PSAC, whose members were largely drawn from elite universities. Historian Charles Maier described PSAC members as being united by

> a common participation in the hermetic ambiance of the Manhattan Project or the MIT Radiation Laboratory . . . they were still ready to presume that the alliance between their long-term research efforts and the aspirations of the United States as a world power was logical, defensible, and meritorious.[72]

Although PSAC, in retrospect, appeared to be a necessary component of the Executive Office of the President, it had taken a dozen years to create such a group. Vannevar Bush's great prominence in science advising for nearly 15 years had obscured the need for a permanent and empowered science advisory structure within the White House, and the influence of the Office of Naval Research during the late 1940s had effectively arrogated the planning role to the Navy. The Soviet Union's atomic and hydrogen bomb detonations and launching of Sputnik had pressured Eisenhower to formalize the science advisory role, leading to the creation of PSAC soon after. Rabi judged PSAC's creation as critical to executive decision-making. He wrote,

> To the best of my knowledge this setup is unique, in that no other government has shown the foresight and understanding to realize what a central role science and technology play in the maintenance and evolution of modern society. . . . We can even be grateful for the beep, beep, beep of Sputnik I, which hastened the decision to introduce these administrative innovations.[73]

Making Engineers

Efforts to trump Soviet engineering were predicated on trumping Soviet engineers, but the Soviets seemed to have a strong lead. The Soviet education system was designed to identify and retain the most talented students and divert them toward scientific and engineering training: in the 1950s, seven of eight Soviet doctoral candidates were earning their degrees in a scientific discipline, as opposed to only four of eight Americans.[74] James Killian wrote in 1954,

> As we face this challenge to American technology, we watch the Soviet's steady growth in technological strength. The Soviet success in achieving an educational output of scientists and engineers greater than our own has brought an awakened realization of the present responsibilities of our own educational system.[75]

To counter this mass of newly emerging scientific and engineering talent, the United States needed to raise both the size of its scientific engineering effort and the quality. Quantity was being addressed. The number of Americans earning a bachelor's degree rose sharply during the postwar years, from approximately 12,000 annually during the war to nearly 50,000 by 1950.[76] Fellowship support from the Office of Naval Research, the AEC, and the NSF were expanding steadily; by 1954, the federal government was supporting 47,000 graduate and postdoctoral students, of which nearly half were in scientific and engineering disciplines.[77]

Raising quality was the trickier challenge, as it rested on raising faculty research productivity. Leaders of elite science and engineering programs around the country repeatedly asserted that the best way, perhaps the *only* way, of training excellent scientists and engineers was to immerse them in a graduate program populated by faculty actively conducting research at the forefront of their disciplines; yet at the time, over 80 percent of all faculty in engineering schools were teaching over 12 credit hours per semester.[78] James Killian, president of MIT, expressed concern that too many American engineers were technicians who lacked the "excellence, breadth, and versatility" of true professional engineers. Elsewhere he wrote, "We must seek quality and excellence even at the expense of quantity . . . above all, our need is for emphasis on superior scientists and engineers."[79] Russian training programs had made the mistake of training their own engineers in overly narrow specializations, thus robbing them off the scientific depth and grasp of fundamental principles critical to a true engineer. "Avoid the easy, narrow rut of excessive expertism," admonished Killian.[80]

Frederick Terman, the dean of engineering at Stanford, concurred strongly with Killian's vision for excellence in science and engineering and expressed skepticism at vocational training and on-the-job training (and even traditional undergraduate education) as paths to quality – both of which led (in his opinion) to overly narrow technicians, unable to think creatively and broadly across disciplines. Celebrating "academic types," he stood in opposition to the American pragmatic tradition. He was skeptical of traditional training institutes, teacher's colleges, and professional programs – all paths (in his view) to narrow vocational thinking and professional mediocrity. The very strength of the nation, in Terman's view, rested on world-class science and engineering doctoral programs hosting "vigorous and virile programs of research." Education, and particularly graduate education, would lack "vitality" in the absence of an ongoing program of basic research. The alternative was eventual defeat across many axes of endeavor. "If we do not maintain clear-cut leadership in both science and technology, we are sooner or later lost," he concluded.[81]

Government research funding tended to support Terman's vision of silos of excellence in graduate study, with the preponderance of research funds through the 1950s flowing to only ten universities. The AEC experimented with funding research partnerships between various regional universities (such as Rochester and Vanderbilt) and the national laboratories, as well as a new training program in nuclear engineering hosted by Oak Ridge.[82] Flagship state universities around the country began to build up research in the physical and engineering sciences during that decade, with the Midwestern schools – Michigan, Wisconsin, Minnesota, and Ohio State – showing particular enthusiasm. The nation had enough money and talent to expand the veil of excellence, although few universities would ever draw equal to MIT, Johns Hopkins, Harvard, the University of Chicago, and Stanford. Excellence begat excellence and tended to draw in on itself through an intellectual centripetal force. In Terman's view, this was the desired national goal.

For both Terman and Killian, executives of two of America's greatest universities, the constant leaking of intellectual talent to industry threatened to undermine graduate programs. Industry paid better and rarely required the doctorate; hungry young graduate students could hardly be blamed for leaving their programs early to work on the same research projects as their graduate advisors at double the pay. Sheer intellectual heft was a finite resource in high demand, yet undereducating such talent was myopic in the nation's standoff with the Soviets. Engineering schools, specifically, were most vulnerable to poaching by industry, "more in danger of serious deterioration than other educational institutions," wrote Killian. While universities were the "natural habitats" for serious scientists, this was less true for engineers whose work was intrinsically more applied. Nonetheless, future competitiveness of the nation depended on universities retaining their share of the best talent.[83]

The United States hemorrhaged scientifically gifted students from a young age. A White House research panel in 1958 concluded that of the most promising (top fifth) of middle school students in the United States, 95 percent entered high school, 90 percent graduated high school, 35 percent graduated college, and 2 percent ultimately earned a doctorate in science or engineering.[84] The problems were many – general cultural skepticism of intellectual pursuits, poor quality secondary science teaching, inadequate financial support for higher education – but part of the problem was a general nationwide ignorance of science. The Soviets, with their vaunted and demanding high school curricula, might not produce the world's most brilliant and innovative scientists, but they did produce a population which was knowledgeable about science and which integrated scientific thinking into day-to-day life. "A man cannot be really educated in a relevant way for the needs of our modern life unless he has an understanding of science," warned Killian. Russia, by contrast, had succeeded in developing a broad "appreciation of the value of science" among its nonscientists.[85]

The problem seemed to start with the nation's inability to produce, or more likely retain, well-trained science teachers. Students inclined toward science were generally uninterested in primary and secondary school teaching, and of those who began teaching careers, many departed for industry within a few years.[86] Many high school science teachers were either incapable of teaching to the level of the most gifted students or too stretched by large teaching loads to be able to design and deliver curricula appropriate to scientifically gifted students. Too many students who might have been capable of careers in science were dissuaded from entering long and rigorous programs of study through general peer disapproval or through an American *weltanschauung*, which elevated success in business over success in scholarship. Killian wrote,

> There has been avoidance, if not evasion, of the intellectual tax which must be paid if our intellectual budget is to be balance. At a time when we are forced to match wits on every front with an alert enemy of growing prowess, America's survival requires that the American people reduce the anti–intellectualism in our schools.[87]

Even President Eisenhower, who rarely embraced the intellectual life, was firm on the point. "Our schools are more important than our Nike batteries, more necessary than our radar warning nets, and more powerful even than the energy of the atom," he stated in a 1957 speech.[88]

One advantage of the Soviets was a cultural respect for scientists that the American lacked. Soviet scientists were granted stature and authority in their country. They were well paid (by Soviet standards) and received priority in housing, access to travel, and privileges in specialized food stores. Young Soviet scientists were awarded lifetime stipends, bonuses for mentoring doctoral students and publishing scientific articles, and positions of authority within the party. A career in science was a pathway to elite status within the country, with special privileges and a better quality of life. By contrast, American academic scientists found themselves fighting for grants, consigned to middle-class pay, and culturally dismissed as overly cerebral and aloof. While there was substantial honor for the scientist *within* the profession, there was precious little outside the profession. Scientists were rarely promoted to executive positions within companies, rarely gained positions of power and authority within government, rarely ran for office, and were rarely rewarded the respect garnered by "public" intellectuals. A 1958 White House panel articulated a different vision for the talented scientist with rare intellectual and creative gifts which were critical to the long-term success of the nation. "The conservation and development of this most valuable of our resources must be one of our primary concerns," the panel concluded.

> It is of superlative importance that such capabilities be identified at the earliest possible age, that they receive the best possible counseling, that suitable environment be provided for their full development and, when they are ready, for their unrestrained search for new knowledge.[89]

Loyalty

Questions of secrecy and loyalty plagued military research during the Cold War. Much of the funded research had military applications, and the government reasonably wished to ensure that sensitive data would remain secret. At the same time, the government funded a significant amount of basic research neither related to national security nor immediately applicable to weapons development or defensive technology. But where to draw the line? Scientific administrators, members of Congress, and political leaders debated the optimum level of secrecy – a level adequate to safeguard research directly related to national security without actually hampering the free-flow of information and discovery upon which scientific progress was based.

More ominous were debates over loyalty and clearances. Should the government require scientists to swear an oath of allegiance to the United States before receiving research funds, or were existing laws against treason adequate to ensure adherence to confidentiality? Could membership in the Communist party, which was legal in the United States, be grounds to deny a scientist access to funds and classified data? What about a youthful dalliance with leftist politics? Shields Warren, the director of the Division of Biology and Medicine of the AEC, warned in a 1949 memo that

> a number of outstanding creative thinkers go through a "radical" phase in their youth and there is always danger that liberalism or honest criticism of government will be confused with disloyalty, thus keeping some of the best brains out of the program.[90]

During World War II, all branches of the military zealously guarded information about troop and ship movements, armament strength, and emerging technology from the axis powers. After the war, this level of heightened security was quickly turned toward the Soviet Union. Tales of espionage, of "turned" senior American civil servants, and of Soviet infiltration of American labs all served to reinforce American commitments to security. News that Robert Oppenheimer's brother, Frank, had been a member of the Communist party in the 1930s raised the public's fear level, as did the extensive coverage given to the atomic espionage trial of Ethel and Julius Rosenberg in 1951 and their subsequent execution in 1953.[91]

In 1949, the Senate Appropriations Committee amended its annual budget bill with a requirement that no appropriation for the AEC could be used to fund a fellowship for anyone who was a member of an organization advocated for the overthrow of the US government. That requirement seemed reasonable enough, but more worrisome was the addendum that the AEC should refuse to fund any person for whom reasonable grounds existed to believe that he or she was disloyal to the United States.[92] This standard was muddier and invited abuse. Wernher von Braun reported, for example, that he was advised against joining one of the nation's elite scientific organizations, the American Association for the Advancement of Science, because the Army believed that it had been infiltrated by Communists.[93] The Army had even named Communist members including the

organization's president, Harlow Shapley, "a champion of many causes identified as Communist."[94] Fear at times rose to paranoia as American corporations and universities imposed internal loyalty tests on employees, even
when government-funded research was nonclassified and when no such
test was required. In 1949, Detlev Bronk, the chairman of the National
Research Council, warned Oswald Veblen, the president of the Institute for
Advanced Study in Princeton, that loyalty oaths had become the standard
for any organization seeking federal funds, particularly from the AEC or
the DoD. "We were told in no uncertain terms that the oath of loyalty was
a minimum requirement for the continuation of the fellowship program,"
Bronk wrote.[95] Reports of loyalty tests in private contracting firms circulated, even those not engaged in classified research.[96]

Some scientists and scientific administrators pushed back. Requirements
of loyalty oaths or disqualification of loyal citizens who espoused leftist politics threatened to sideline many able scientists and reputable firms. The
AEC officially adopted a policy of not requiring security clearance from any
grantee unless he or she would have access to restricted data or would work
in installations in which such data circulated.[97] L. W. Nordheim, a professor of physics at Duke University and the former director of physics at the
Clinton National Laboratory, for example, protested against the blackballing
of a graduate student at the University of North Carolina for having been
a member of the Communist party, despite the fact that the student did
no work on classified projects. Writing to Senator Clyde Hoey (D., NC),
Nordheim warned that,

> the encroachment of restrictions on science for the sake of security has
> hampered us enough already . . . what would we gain or lose then, by
> extending such restrictions to fields in which no question of security
> is involved? . . . let us preserve our sense of proportion, let us not be
> guided by fear.[98]

Shields Warren warned, more generally, "Science will not long thrive in
secret because intellectual restraints and taboos hamper the type of probing
and wide-ranging thought essential to scientific discovery."[99]

If granting access to law-abiding American citizens was hard, granting
access to well-intentioned foreigners was even harder, and sharing research
discoveries with Soviet scientists was harder still. Scientific insights and
discoveries had always been shared across borders. International scientific
meetings were common as were multinational editorial boards. Nonetheless, funded scientists in the 1950s were generally discouraged from coau-
thoring with foreign authors (even when those authors were from nations
allied with the United States) and at times even data could not be shared
across borders. While the United States had shared some of its progress
in producing sustainable fission with the United Kingdom during World
War II, it largely refrained from cooperating with British scientists in the

years after. But secrecy created its own hazards. It invited hostile response, undermined trust, and raised tension. Edward Condon, a physicist who had worked on the Manhattan Project and who later became the director of the National Bureau of Standards, argued strongly for the interdependent nature of scientific research and for generally encouraging scientific cooperation across borders. He warned of creating an "air of suspicion and mistrust" that could only escalate tension across borders, and he encouraged Americans to "chase this isolationist, chauvinist poison from our minds."[100] Historian Jessica Wang writes of the era that "Secrecy constituted a mind-set that created a false sense of security that defense of the United States could be maintained by the preservation of the atomic monopoly."[101]

In 1949, the State Department's International Science Steering Committee, under the leadership of Lloyd Berkner, issued a report titled "Science and Foreign Relations," which recommended that the State Department set up a science office headed by a science advisor and create positions of scientific attachés in 15 foreign (noncommunist) countries. The attachés and the office would work to create scientific cooperation across borders, organize conferences, disburse scientific publications, and establish scientific exchanges.[102] Four attaché positions were soon created in Stockholm, Paris, Tokyo, and London, although a 1955 report by the Hoover Commission recommended that this function be brought under the supervision of the CIA.[103] The program withered in the late 1950s when the State Department had difficulty identifying and appointing appropriate individuals to the posts. By 1956, a science reporter would describe the program as "ground to a halt."[104]

One fear of international cooperation was that such a program might actually exacerbate tensions between the Americans and the Soviets. The Soviet Union, acutely sensitive to external threats, viewed American alliances through a defensive prism. The American alliance with the United Kingdom had created an asymmetry among allied forces during the waning days of World War II. When the United States, Canada, and most western European nations created the NATO alliance in 1949, the USSR and its Communist allies countered by signing the Warsaw Pact in 1955. The American military presence in West Germany was balanced by a Soviet military presence in East Germany. American support for newly liberated former colonies in the Middle East, Africa, and Southeast Asia was met with comparable Soviet support in those areas. As early as 1945, Dean Acheson, then the Under Secretary of State, warned that evidence of cooperation between the United States, Canada, and the United Kingdom on nuclear physics would "appear to the Soviet Union to be unanswerable evidence of an Anglo-American combination against them."[105]

Scientific cooperation among western aligned nations during the 1950s and 1960s led to the odd phenomenon of parallel scientific enterprises: an American-led coalition and a Soviet-led one. Both enterprises produced excellent work across a variety of disciplines: the Americans had particular success in the biomedical and clinical sciences and in experimental physics, while the Soviets maintained parity in pure math, theoretical physics, and some areas of aerospace engineering. Both enterprises relied on peer review, meritocracy, academic appointments, and scientific publications to move science along; yet few research projects crossed the barrier. Not until efforts at détente in the 1970s, marked most notably by the Apollo–Soyuz joint docking maneuvers in 1975, would lines of communication begin to open.

While the issue of parity and competitiveness with the Soviets dominated American science policy all through the three decades following the end of World War II, nowhere can it be seen as nakedly as in the development of the hydrogen bomb and in the co-synchronous concern about the quality and quantity of American scientific and engineering talent. Discussion surrounding the development of the hydrogen bomb almost entirely ignored military tactics and strategy and focused, instead, on a metaphysical conception of cultural and intellectual superiority. The society which was more robust *in its essence* was the one capable of producing the smartest scientists who could direct the most talented engineers into building the most sophisticated possible weapon. The fact that that weapon could never be used was immaterial. The mere production of the weapon signaled to the opposing regime that its possessor would ultimately triumph. The Cold War had moved into an almost spiritual realm in which culture, knowledge, expertise, and sheer scientific brilliance would be the currency moving forward.

Notes

1 An enduring puzzle is why the US defense establishment largely underestimated Soviet capability in nuclear science in the 1940s. Karl Compton hypothesized that the USSR was able to build on German and Czechoslovakian expertise when it effectively integrated those two nations into its sphere of influence and control after the World War II armistice. See Compton to Bush, February 11, 1953, VB, 609/ Karl Compton.

2 "A Report to the President, Pursuant to the President's Directive of January 31, 1950," April 7, 1950, PSF, 174: Atomic Energy Advisory Committee, p. 1.

3 "The Key Role of Basic Research in Maintaining the Technological Position of the U.S. in the World Today," 1958, OSAST, 1: Basic Research, 12/57–11/60.

4 Waterman, "Russian Science Threatens the West," *Nation's Business*, September, 1954.

5 Bevis to Eisenhower, November 26, 1957, DDE, PSAC:4, Committee on Scientists.

6 Ibid., p. 2.

7 James Killian, "The Challenge of Soviet Science," April 8, 1957, MIT, 195: SAC, p. 6.

8 National Science Foundation, "General Considerations Concerning United States Progress in Science and Education," July 31, 1958, AW, 28: NSF Policy Papers, p. 2.

9 Lloyd Berkner, "Wanted: A National Science Policy," *The Atlantic Monthly*, 201, January 1958, p. 40.

10 W. Sterling Cole, "Science and Statecraft," *Science*, 121, June 24, 1955, p. 886.

11 "Efficiency and Results in U.S. Military Technology," OSAST, 1: DOD, 1957.

12 "A Report to the President, Pursuant to the President's Directive of January 31, 1950," April 7, 1950, PSF, 174: Atomic Energy Advisory Committee, p. 9.

13 Henry Cabot Lodge, Jr., "Soviet Air Power Has Big Edge Over U.S.," April 30, 1951, VB: 1403, Henry Cabot Lodge, Jr., p. 2.

14 James Killian, "Survival in an Age of Technological Contest: A Preliminary Outline for Strengthening Military Research and Development," October 15, 1957, MIT, 195: SAC.

15 Ibid., p. 11.

16 Killian, "The Challenge of Soviet Science," p. 7.

17 Ibid., p. 8.

18 PSAC, "Report of the Computer Panel," September 11, 1959, OSAST, 1: Computers, 1/59–6/60.

19 Vannevar Bush, "The Weapons We Need for Freedom," 1952?, VB, 3091.

20 For more, see the comprehensive "Report of the Research Panel," of PSAC, June 17, 1958, OSAST, 1: Basic Research, 12/57–11/60.

21 McMahon to Truman, November 21, 1949, SRF, 11A:1/2.

22 "Report to the President," April 7, 1950, PSF, 174: Atomic Energy Advisory Committee, p. 12.

23 Henry Jackson, Speech to the House of Representatives, October 9, 1951, IIR, 17:7, p. 2.

24 "Memorandum for the President," January 17, 1952, SRF, 11A:5.

25 Brien McMahon, "Address to the Senate," September 10, 1951, EOL papers, 108631.

26 "1951 – The Payoff Year," *Business Week*, July 28, 1951, pp. 3–12.

27 "Report to the President by the Special Committee of the National Security Council on the Proposed Acceleration of the Atomic Energy Program," October 10, 1949, SRF, 11A:5, p. 5.

28 We will never know. Using information on Japanese military readiness gained after the war, allied estimates for total invasion fatalities dropped significantly. But, in considering whether or not to use the atomic bomb, Truman based his decision on his military advisors' estimate of 2,500,000 *total* fatalities (including Japanese). There is substantial literature on the ethical parameters of the decision. For a pithy and compelling case, see Michael Walzer and Paul Fussell, "An Exchange on Hiroshima: On the Moral Calculus on the Bomb," *New Republic*, 185, August 1981, p. 13.

29 Conant to Bush, September 27, 1945, VB, 608: "James Conant."

30 Arthur Compton to A. J. McCartney, March 18, 1946, VB, 608: "Arthur Compton," p. 1.

31 Ibid., p. 3.

32 Jon Bugher, "Delayed Radiation Effects at Hiroshima and Nagasaki: A Progress Report," DB, FA 965, 28: AEC, 1952, p. 1.

33 Ibid., p. 10.

34 AEC, Press Release: "Atomic Bomb Casualty Commission to Continue Studies of Japanese Atomic Bomb Survivors," June 18, 1950, DB, FA 965, 28: AEC, 1950. Radiation exposure was both more and less harmful than initially supposed. By 1950, researchers had identified infection, anemia, hemophilia, and problems with ionic exchange as effects of radiation exposure. However, survivors of the Japanese bombings exhibited a high degree of variation. Clothing, the presence of walls, distance, and even physical position at the time of the blast, all significantly affected sickness and recovery. See Shields Warren, "Statement to the Joint Congressional Committee on Atomic Energy," March 17, 1950, DB, FA 965, 28: AEC, 1950.

35 Vannevar Bush, *Modern Arms and Free Men*, MIT Press, 1949, p. 101.

36 Ibid., p. 104. Bush became somewhat of a strategic military visionary after the war. In the same treatise he envisioned bomb delivery through submarine-launched guided missiles and that those submarines which would carry those missiles would be powered by shipboard nuclear reactors. Such a vision ultimately proved true with the launching of the nation's first nuclear-powered missile-carrying submarine, the USS George Washington, in 1959.

37 Borst to Truman, December 5, 1951, PSF 96: Atomic Bomb.
38 The text of the speech is widely available. See www.ourdocuments.gov/doc.php?flash=false&doc=90&page=transcript.
39 Kistiakowsky to PSAC, date unknown, DB, FA 965, 15: Kistiakowsky, G.
40 Khrushchev to Eisenhower, July 2, 1958, DB, FA 965, 15: PSAC. The premier also requested that the USSR be given the right to conduct aerial inspections over east Asian territory and even parts of Alaska.
41 In fact, Oppenheimer had raised concern about the health effects (and necessity) of atomic bomb tests as early as 1946. See Oppenheimer to Truman, May 3, 1946, SRF, 11A:5. It was difficult to discern the true effects, though. Shields Warren, the director of the Division of Biology and Medicine of the AEC suggested in a public hearing that waters around the Bikini testing site had only "trace quantities" of radiation. See Warren, "Statement to Joint Congressional Committee on Atomic Energy," March 17, 1950, DB, FA 965, 28: AEC, 1950.
42 General Advisory Committee, May 7, 1958. OSAST, 1: Atomic Energy, 11/57–12/60.
43 Another barrier to continuing test ban discussions that summer was the outbreak of civil war in the newly independent Congo, which quickly evolved into a proxy war between the two superpowers. For more on the history of a comprehensive test ban, see Glenn Seaborg, *Kennedy, Khrushchev, and the Test Ban*, University of California Press, 1981, particularly chapters 1 and 2.
44 Johnson to Bundy, July 11, 1963, LBJ, NSF – Charles Johnson, 30: Nuclear Atmospheric Tests.
45 Quoted in Herbert York, *The Advisors: Oppenheimer, Teller and the Super Bomb*, W. J. Freeman, 1976, p. 47.
46 Hans Bethe, Peter Kihss, and William Kaufmann, "The H-Bomb and World Order," *Foreign Policy Reports*, 26:8, September 1, 1950, p. 83.
47 Quoted in James Hershberg, *James B. Conant: Harvard to Hiroshima and the Making of the Nuclear Age*, Alfred Knopf, 1993, p. 473.
48 Quoted in Annie Jacobsen, *The Pentagon's Brain: An Uncensored History of DARPA*, Little, Brown and Company, 2015, p. 13.
49 "A Report to the President, Pursuant to the President's Directive of January 31, 1950," April 7, 1950, PSF, 174: Atomic Energy Advisory Committee, p. 3.
50 McMahon to Truman, November 21, 1949, SRF, 11A:1/2, p. 3.
51 Strauss to Truman, November 25, 1949, SRF, 11A:12, p. 1.
52 Ibid.
53 Jacobsen, *The Pentagon's Brain*, p. 13.
54 Quoted in York, *The Advisors*, p. 52.
55 Ibid., p. 53.
56 Quoted in Ibid., p. 92.
57 Ibid., p. 92.
58 Bush to Conant, March 29, 1954, VB:614, p. 2.
59 Ibid., p. 4.
60 Charles Maier, "Introduction," in George Kistiakowsky, *A Scientist at the White House*, Harvard University Press, 1976, pp. lv–lvi.
61 Press Release, Executive Office of the President, May 27, 1952, IIR, 43:3.
62 Buckley to Truman, May 1, 1952, MIT, 194: SAC, p. 3.
63 See Jerome Wiesner, *Where Science and Politics Meet*, McGraw-Hill, 1961, pp. 44–45. The Gaither Panel made recommendations on strengthening US defensive systems against nuclear attack. The Killian Panel focused on neutralizing Soviet air threats.
64 See Audra Wolfe, *Competing with the Soviets*, Johns Hopkins, 2013, p. 49.
65 I.I. Rabi, "Science and Public Policy," Compton Lecture No. 2, March 8, 1962, IIR, 65:2, p. 13.

66 "Background Statement of the Activities of the President's Science Advisory Committee," 1957–1959, IIR, 46:3.
67 DuBridge to Killian, December 16, 1955, MIT, 195: SAC, Technological Capabilities Panel, pp. 1–2.
68 "A Program for the Federal Government to Help in Releasing the Full Scientific Capacities of the U.S. and Her Allies," IIR, 45:6, p. 7.
69 Commission on Organization of the Executive Branch of the Government (Hoover Commission), *Research and Development in the Government: A Report to the Congress*, GPO, Washington, DC, May 1955, p. 10., as quoted in Ibid., p. 11.
70 Quoted in Wolfe, *Competing with the Soviets*, p. 50.
71 Ibid.
72 Maier, "Introduction," in *A Scientist at the White House*, p. lvii.
73 I.I. Rabi, *My Life and Times as a Physicist*, Claremont College Press, 1960, pp. 27–28.
74 Scientific Advising Research Panel, "The Key Role of Basic Research in Maintaining the Technological Position of the U.S. in the World Today," OSAST, 1: Basic Research, 12/57–11/60.
75 James Killian, "America's Capacity to Maintain Technological Leadership," January 23, 1956, DB, FA 965, 15: PSAC ODM, 1954, p. 2.
76 Buckley to SAC, January 15, 1952, "Engineering Achievement Potential," MIT, 194: SAC, addendum: figure 2.
77 Alan Waterman, "Role of the Government in Science Education," *Scientific Monthly*, 82:6, June 1956, p. 288.
78 Killian, "America's Capacity to Maintain Technological Leadership," p. 6.
79 Ibid., p. 4.
80 James Killian, "The Challenge of Soviet Science," April 8, 1957, MIT, 195: SAC, p. 10.
81 Frederic Terman to Members of the PSAC Panel on Basic Research and Graduate Education, draft of a policy brief, January 8, 1960, OSAST, 13: Basic Research, pp. 4–6.
82 See Gordon Dean, "Remarks to the Symposium of the American Medical Association," June 26, 1950, DB, FA 965, 28: AEC, 1950.
83 Killian, "America's Capacity to Maintain Technological Leadership," p. 5.
84 Alan Waterman, "Russian Science Threatens the West," *Nation's Business*, September 1954.
85 Killian, "The Challenge of Soviet Science," p. 10.
86 Waterman, "Russian Science Threatens the West," p. 4.
87 Killian, "America's Capacity to Maintain Technological Leadership," p. 5.
88 Quoted in Killian, "The Challenge of Soviet Science," p. 11.
89 Scientific Advising Research Panel, "The Key Role of Basic Research in Maintaining the Technological Position of the U.S. in the World Today," OSAST, 1: Basic Research, 12/57–11/60, p. 9.
90 Shields Warren, "Statement Relative to Fellowship Program," September 27, 1949, DB, FA 965, 28: AEC 1949, p. 2.
91 James Walter, "Frank Oppenheimer Was at Oak Ridge, Los Alamos Plants," *Washington Times-Herald*, 1947.
92 Zook to Bronk, July 13, 1949, DB, FA 965, 28: AEC 1949.
93 The AAAS was on the Army's "pink list" of suspicious organizations. See Wayne Biddle, "AAAS's Very Own Red Scare," *Science*, 260:5107, April 23, 1993, p. 486.
94 Ibid. Shapley's son denied that he was ever a party member.
95 Bronk to Veblen, June 25, 1949, DB, FA 965, 28: AEC 1949.
96 See R.W. Stoughton to Brien McMahon, May 25, 1949, DB, FA 965, 28: AEC 1949.
97 Carroll Wilson to Bronk, October 8, 1948, DB, FA 965, 28: AEC 1949.

98 Nordheim to Hoey, May 16, 1949, DB, FA 965, 28: AEC 1949.
99 Warren, "Statement Relative to Fellowship Program," p. 2.
100 Quoted in Jessica Wang, *American Science in the Age of Anxiety*, University of North Carolina Press, 1999, p. 22.
101 Ibid.
102 Lloyd Berkner, *Science and Foreign Relations*, United States Department of State, 1950.
103 Wilson Lexow, "The Science Attaché Program," in *Studies in Intelligence*, vol. 10, Central Intelligence Agency, Spring, 1966, p. 23.
104 "What Happened to Science in State?" *Chemical and Engineering News*, January 9, 1956, p. 112.
105 Acheson, "U.S. Policy Regarding Secrecy of Scientific Knowledge About Atomic Bomb and Atomic Energy," September 25, 1945, PSF, 174: Atomic Bomb: Cabinet: James Byrnes.

8 Power, Plowshare, and Peaceful Atoms

Atoms for Electricity

In January 1959, Robert McKinney, the US Ambassador to the International Atomic Energy Agency (IAEA), proposed a panel report to study the prospects of commercially generating electricity through atomic power. McKinney, a wealthy investor from Oklahoma, had served as President Eisenhower's ambassador to Switzerland and as Assistant Secretary of the Interior before taking on his role at the IAEA. An inveterate skeptic, his expectations of the panel were limited; he suspected that commercial use of atomic power would be uneconomical and uncompetitive. The high cost of reactor construction, the many material challenges of working in a highly radioactive environment, and the continued high cost of uranium enrichment all worked against the economic competitiveness of atomic power in a world with increasing known deposits of fossil fuels. In addition, the public's association of atomic power with horrific destruction and radiation poisoning worked against the future of civilian nuclear power. Public resistance would only increase with local debates over reactor construction and the inevitable atomic accidents.[1]

Particle physicists understood the potential to use fission for civilian power generation from the beginning of the atomic age. Enrico Fermi had reportedly calculated the equivalent energy in tons of coal of a kilogram of enriched uranium on the day he first achieved sustained fission in 1942. The engineers who designed the Hanford reactor during World War II sited it on the Columbia River, understanding that it would produce so much heat during normal operations that it would require 75,000 gallons of water to cool it each minute. It was a short leap to recognize that if the heat could be drawn off in a controlled manner, it could be used to produce steam and power turbines. The mechanics were established and understood; only the heat transfer process stood as a barrier.

Transferring heat out of a nuclear pile to a heat exchanger was challenging, however. Any coolant flowing through the active pile would become highly radioactive while at the same time drawing off neutrons which were necessary in perpetuating the reaction. The high temperatures (approaching 10,000 degrees) and intense pressure (sufficient to blow up the reactor)

DOI: 10.4324/9781003363897-8

would require sophisticated and expensive metallurgical solutions. At the same time, producing enriched uranium was itself an energy-intensive proposition. The gaseous diffusion process of enrichment, the primary method in use in the late 1940s, required immense amounts of power and a huge infrastructure; the structure that housed the Oak Ridge enrichment machinery was the world's largest building at the time of its construction in 1943. Moreover, the power produced by a civilian reactor would need to compete on the commercial market with power produced by coal and oil plants, which were easy to build and maintain and capable of exploiting the world's growing known reserves of fossil fuels.

Few in the field of nuclear physics found the idea particularly compelling. Developing civilian power generation was fundamentally an engineering challenge rather than a scientific one – that of working within certain parameters to produce economically competitive power. Eugene Wigner, one of the group of Hungarian physicists who had worked on the Manhattan Project, testified to the Joint Committee on Atomic Energy that he knew of few people "willing to devote the next twenty years of their lives to the problem of economic nuclear power."[2] The national laboratory structure tended to reward more basic research oriented toward military applications; price rarely entered into their considerations. Wigner told the committee, "It is not in the nature of government institutions to give economic considerations a dominant role."[3]

Electricity generated through nuclear power, while not economically competitive for the United States, might prove to be a useful alternative for nations lacking their own fossil fuel deposits or for isolated regions in which shipments of large quantities of fossil fuel were impractical. In the future, improved reactor and enrichment technology might reduce the cost of atomically generated electricity to a price competitive with that produced in conventional plants. Boosters of the technology tended to understate the technical challenges, though. In 1947, editors at *Business Week* predicted that economically competitive nuclear power generation was only five years away, and in 1962, the *New York Times* optimistically predicted that nuclear power plants were already "on the verge" of being competitive with traditional coal-powered plants.[4] In truth, these predictions failed to account for growth in known coal and oil reserves, declining extraction costs, stagnant progress on gaseous diffusion technology, and general security concerns about distributing enriched uranium fuel to civilian power companies. An internal White House memo stated in 1961 that the need for a nuclear power plant was not imminent, "except in the prestige sense."[5] Nonetheless, on March 17, 1962, President John F. Kennedy requested that the AEC "take a new and hard look at the role of nuclear power in our economy" and identify potential objectives for meeting the nation's energy needs with nuclear power generation.[6]

Private industry responded rapidly. General Electric, already involved in nuclear power generation through its management of the Hanford and

Knolls laboratories, viewed any increase in electrical power generation as good for business. Monsanto, the general contractor in charge of the Oak Ridge laboratory, took a more cautious approach but generally sided with investing in civilian nuclear power production.[7] Both companies had been experimenting with different reactor designs since 1947, and both were aware of the growing efficiency of the piles and the increasing competitiveness of nuclear-generated electric power. In 1963, Robert Wilson, a Cornell University physicist who was serving as a member of the AEC, estimated that as soon as 1970, a nuclear power plant would be able to generate electric power at a price competitive with coal – about 5.6 millicents (1/1000 of a US dollar) per kilowatt hour (kwh) or roughly the prevailing price of coal-generated electricity.[8]

Jersey Central Power and Light Company (JCP&L) coupled with the Niagara Mohawk Power Corporation the following year to construct two large nuclear power plants, expecting to save about 3 millicents per kilowatt hour over power generated in a conventional plant. JCP&L's initial $60 million reactor was ordered from General Electric for installation at Oyster Creek and was slated to produce 515 kilowatts of power annually for 30 years.[9] A slight innovation in fuel rod design – using uranium dioxide rather than metallic uranium – lowered operating costs for the reactors. The announcement of the contract drew the attention of firms involved with atomic energy production. One trade publication described the "historic turning point" in moving toward nuclear power and envisioned nuclear plants competing for up to a third of all new power contracts in the coming year.[10] General Electric predicted that within two years a fifth of all new power plants constructed each year would be nuclear. Later that year the Westinghouse Corporation announced that it would compete for nuclear power plant contracts of up to one million kilowatts. President Lyndon Johnson, perhaps prematurely, announced that civilian nuclear power generation was now an "integral part of the American industrial scene" and predicted a "steadily increasing share" of the economy centered on nuclear energy.[11]

Orders for nuclear power plants increased sharply over the next two years. By 1967, construction contracts for nuclear power plants amounted to 57 megawatts or 15 percent of all of the country's generating capacity. General Electric, Westinghouse, Combustion Engineering, and Babcock and Wilcox were under contract to build 90 new power plants, including 15 for foreign customers.[12] The pace of power plant construction was practically doubling each year. The AEC estimated that by 1980, nuclear power plants would generate between 80 and 110 megawatts of power or a third or all US electric power.[13]

Notably, none of the commercial reactors under construction in the late 1960s were breeder reactors – those designed to convert nonfissionable uranium into plutonium (Pu^{239}), thorium (Th^{232}), or a fissionable isotope of uranium (U^{233}). Mined uranium was only 0.7 percent U^{235}, the common fissionable isotope of uranium. This was enriched to approximately 3 percent U^{235} through either a gaseous diffusion or a gaseous centrifuge process, which was then fed into a reactor in the form of fuel rods. (Weapons-grade uranium needed to be enriched to over 80% U^{235}.) This meant that most reactors were failing to use 99.3 percent of the mined uranium. But in a reactor designed for breeding, extra neutrons were captured by the U^{238}, converting it to fissionable metals and isotopes allowing almost 70 percent of all of the uranium to undergo fission and produce power. As a result of the process, a given mass of uranium could generate nearly 100 times as much power as was possible in the absence of breeding. The problem, however, was that the process created plutonium – a fissionable product which could easily be used in constructing nuclear warheads. The prospect of theft of plutonium, or of covert sales of plutonium to enemy nations or even terrorist organizations, worried military planners. In an effort to thwart nuclear proliferation, the AEC carefully limited civilian construction to nonbreeder reactors.

Even within its own laboratories the AEC was leery of breeding plutonium. In addition to its original plutonium breeder at Hanford, Washington, it possessed a breeder reactor in Idaho Falls, Idaho; it also planned to build a third breeder at the Argonne National Laboratory outside of Chicago. In the early 1960s, AEC commissioners and White House science advisors debated building a new dual-use Hanford pile, capable of generating 900 megawatts of power while producing plutonium.[14] The reactor would be the largest in the world – nearly four times larger than the largest reactor then operating in the United States, and eight times larger than the largest Soviet reactor. President Kennedy suggested, without irony, that the machine would be a "dramatic demonstration to the world of the peacetime application of the atom," given that it would have economic value even if a comprehensive disarmament treaty could be negotiated.[15] Pressure toward building commercial breeder reactors began to subside by the mid-1960s, however, as substantial discoveries of uranium deposits around the world and more efficient centrifuge technology diminished the cost of conventional nuclear fuel. By May 1964, the AEC announced that it would be mothballing its existing breeder reactors over the next two years.[16]

Proliferation of nuclear weapons was a constant concern as civilian power generation increased. Under the Atomic Energy Act, the AEC owned all enriched radioactive material in the country. No private firm had the capacity to enrich its own uranium fuel, and the AEC guarded its monopoly assiduously. Starting in 1961, however, the AEC began to lease enriched uranium to private firms for use in both experimental and commercial reactors, with the understanding that all depleted uranium (and adhering metals and isotopes) would be returned to the AEC. Through the 1950s, when virtually all nongovernment reactors were designed for research, the burden on the AEC to produce enriched uranium was modest. As commercial nuclear power grew during the 1960s, however, the AEC was forced to devote greater resources to enriching and distributing fuel. It raised prices several times during the decade in an effort to recapture some of its costs and by 1964, agreed to actually sell (rather than lease) material to conventional power plants.[17] The first international sale of enriched fuel took place in 1969 to the Swedish firm OKG for a new power plant to open in Sweden the following year at prices calibrated to make a small return for the government.[18] Charles Schultze, the director of the Bureau of the Budget, informed the President that the sale of enriched uranium was a commercial activity entailing the usual economic risks, for which the government should get a "business rate of return – higher than 5%."[19]

Even amid such precautionary measures, fear of nuclear proliferation was omnipresent. Advances in centrifuge technology made the gas centrifuge enrichment process more efficient and more accessible. (Gaseous *diffusion* plants, by contrast, were so immense and expensive that no private firm or small country could conceivably construct one in secret.) Commercial engineering firms actually undermined the government's efforts at securing the nuclear stockpile as they produced more sophisticated centrifuge designs during this time. The centrifuges were getting thinner, faster, more efficient, and more durable. Thermo-convection design (heating the bottom of the centrifuge) facilitated actually removing the enriched uranium hexafluoride gas from the vessel. Glenn Seaborg warned Walter Rostow, President Kennedy's National Security Advisor, of growing commercial markets for the enriched gas and of the proliferation of centrifuge technology which would make control of enriched uranium more difficult. Seaborg wrote, "We believe that such activities could have the effect of stimulating further interest abroad in the gas centrifuge process, thus leading to a proliferation of capabilities to enrich uranium for weapons" and recommended that the nation prohibit privately sponsored work on centrifuge development.[20] That is, for all of its caution, nuclear technology was evading the AEC's control and would likely continue to do so in the future.

Civilian nuclear power generation was both more successful and more dangerous than many had envisioned at the beginning of the decade. Nuclear-generated electric power was competitive with coal-generated power by the mid-1960s and getting cheaper as reactor designs improved and uranium ore prices declined. Despite the public's association of nuclear power with wartime destruction and radiation sickness, there had been generally little popular opposition to the development of the plants. At least five American companies were taking on construction contracts, and multiple power companies were planning to expand into the sector. At the same time, the AEC's tight control of the enriched fuel which powered the reactors was proving problematic. The AEC had been designed as a weapons and research agency; not as a fuel provider for the civilian energy sector. Its decision in 1967 to sell, rather than lease, material relieved it of some of the holding costs of maintaining such a large uranium stockpile but still left it responsible for tracking all fissile material and accounting for post-fission products. And developments in reactor and centrifuge technology were making the fission process more difficult to control. A half dozen nations had civilian nuclear reactors by 1970, and more nations were considering adopting the technology. The technology was escaping American control.

Radioisotopes

From the early days of the MED, scientists identified radioactive isotopes of common elements produced in reactors and accelerators. These isotopes – a variant of an ordinary element such as carbon, phosphorous, iodine, or strontium – had the same numbers of protons in their nuclei but emitted radiation in the form of alpha, beta, or gamma rays. The intensity and type of radiation were specific to the element, as was the half-life of the radio-isotope: that is, the amount of time during which it would continue to emit radiation.[21]

From an early point, MED scientists recognized that radioisotopes might be used in medicine. Most obviously, they could be used for diagnosis and mapping. Different elements and molecules took different paths through the body and became concentrated in different organs or in different cellular organelles. Muscles absorbed glucose during metabolism; for example, while neurons took in potassium; red blood cells bound with iron; plants cells fixed nitrogen; and the thyroid attracted iodine. If radioactive isotopes of these elements, or of constituent elements in these molecules, were introduced into various plant and animal bodies, their distribution could be tracked on an X-ray and used to identify abnormalities in cell growth, metabolism, or vasculature.

Equally promising was using the isotopes to selectively irradiate tumors or noncancerous tissue growing in abnormal ways. Paradoxically, radiation both caused and cured certain types of cancer. If radioisotopes which emitted radiation of a certain type and intensity could be absorbed into a tumor, it might kill the tumor and negate the need for surgery. The most obvious potential clinical use was in treating overactive or cancerous thyroid tissue, as thyroid selectively absorbed iodine in a way that few other organs did. But there were almost certainly other forms of cancer which might be treated in this manner.

In 1946, the MED announced a program to distribute the isotopes coming out of its reactors for research use. The first order, delivered on August 2 of that year, was a shipment of one ten-thousandth of an ounce of carbon-14 to the Bernard Free Skin and Cancer Hospital in St. Louis for use in cancer and diabetes research and to track carbon deposits in bones, teeth, and fat. The order was priced at $367, plus $33 for shipping. A War Department press release claimed that the price reflected the cost of production in the Clinton laboratory. Given that the lab had cost hundreds of millions of dollars to build, it was not clear how the government had derived the price.[22] Other early isotope shipments went to researchers at the University of Pennsylvania (for studying sugar metabolism in diabetic animals); the

University of Chicago (glucose distribution in photosynthesis); the University of Minnesota (deposition of carbon compounds in dentin); and the University of California (fat metabolism in the liver).

After 1947, the AEC took over the work of distributing the isotopes without changing the criteria for disbursement. In 1949, the AEC made available the first isotopes produced in a cyclotron (rather than in a reactor) and added radioactive beryllium, sodium, iron, zinc, arsenic, and iodine to its inventory. Some of these isotopes had considerably longer half-lives than the earlier distributed isotopes, making them better candidates for studies over prolonged periods. The sodium-22 coming out of the cyclotrons, for example, had a half-life of three years rather than the 14-hour half-life of sodium-24.[23] The AEC also expanded distribution to 27 nations around the world, although it excluded all Soviet-aligned nations (except Cuba) from the distribution list. New work mapped the distribution of beryllium in different animals, chloride deposits in chicken tissue, and phosphorous in brain tumors.[24] Soil scientists mapped the uptake of phosphorous into plant tissue – work which might possibly help the nation's farms use synthetic fertilizer more efficiently.[25]

By 1951, the AEC had distributed 12,000 shipments of isotopes to 300 institutions in the United States and 150 abroad.[26] AEC commissioner Sumner Pike reported on the "spectacular results" of the research, including radio-iodine treatment of thyroid cancer, radio-phosphorous location of cancers, radio-boron treatment of brain tumors, and radio-hafnium modification of the adrenal glands. There was much promise in using radioactive gallium to treat bone cancer and radio-cobalt, the "workhorse among the radioisotopes", for many areas of the body.[27] Xenon, boron, iodine, gold, and mercury all held promise for future therapeutic applications.[28]

By 1965, celebrating 20 years of the isotope distribution program, Glenn Seaborg cited more than 10,000 state and federal licenses for radioisotope use in medicine, industry, and research. More than 3,000 hospitals routinely used radioisotopes in gauging, radiography, tracer studies, and therapeutics. The food industry used radiation for pasteurization and sterilization. The nation's space program used radiation from isotopes to power small observation equipment and renew batteries. Even the construction industry was using radiation to polymerize plastic-treated wood for durable outdoor construction.[29] Oak Ridge National Laboratory continued to be the primary source for the nation's radioisotopes, but the AEC contracted with university-based cyclotron labs for small amounts of the rarer elements.

The AEC's radioisotope distribution program was central to the broader effort of finding peaceful applications for atomic technology. Although atomic research had been dominated from the start by military considerations, the potential for civilian application had been a stated goal of early congressional and AEC leaders. In the case of the isotopes, civilian applications exceeded expectations. Therapeutic applications were exploited as early as 1949, and the demand for the isotopes grew more rapidly than early administrators of the program had expected. Shields Warren, the director of the AEC's Division of Biology and Medicine, reported to Congress in 1949 that hundreds of scientific papers had already been published using data drawn from research with radioisotopes and that the AEC (to its surprise) had not needed to subsidize research or cajole researchers. "The commission has found that it has simply to produce the isotopes," he reported. "The research workers, clinics, laboratories, and hospitals need no encouragement to put them to work."[30]

Transportation and Desalination

The success of using nuclear reactors for electric power production gave rise to further applications of nuclear power. One of the most obvious uses was in transportation, where mobile nuclear reactors could heat water to move turbines and power craft forward. In 1955, President Eisenhower approved development of an experimental nuclear-powered merchant ship, the *Savannah*, named after an earlier ship of the same name that had been the first to cross the Atlantic under steam power in 1819.[31] The new *Savannah* was built in 1957 in Camden, NJ, and christened in 1959 by the First Lady, Mamie Eisenhower. It sailed to Yorktown, Virginia under conventional diesel power where a nuclear reactor was installed. It was fully functional under nuclear power by 1961.

The ship was designed to demonstrate the peaceful uses of atomic energy and as such was not designed to optimize freight hauling. It initially carried up to 100 passengers alongside its reactor and 8,500 tons of commercial freight. In 1965, it ceased carrying passengers, but it continued to haul freight until 1971, when it was decommissioned. By that time *Savannah* had sailed 450,000 miles and had been refueled once.

The *Savannah* was never an economical venture. With diesel fuel selling for $20 per ton in 1960, the cost of nuclear fuel alone was nearly four times the cost of conventional diesel fuel. Moreover, the ship carried roughly twice the crew as a conventional freighter (including nuclear engineers), and its crew members required more extensive and specialized training. Several countries around the world, including Australia, New Zealand, and Japan, refused to grant the ship docking privileges, and the ship's freight hold, designed around the central reactor space, was not accessible by the more automated port cranes. *Savannah* also required a specialized nuclear tender ship to service her (the *Atomic Servant*, designed to remove radioactive wastewater and core material), adding to the unusual costs of her operations. The *Savannah* was actually owned by the federal government's Maritime Administration, which leased it to the American Export Lines for operation.

Only four nuclear-powered cargo ships were ever built – too few to create a robust nuclear servicing infrastructure. The economics were substantially different from military uses of nuclear power, in which economic concerns were subordinated to combat concerns. Naval development of nuclear propulsion in aircraft carriers and submarines was driven entirely by tactical considerations, although as more of these ships and submarines were built operating costs per mile declined substantially. The *Savannah* is best seen as an experiment in alternative civilian uses of nuclear power and as a demonstration of American technological prowess within the context of the Cold War. It failed on the first axis and was ultimately superseded on the second by the visible successes of the American nuclear naval fleet.

Nuclear power could not compete with diesel power for maritime use. But nuclear power could easily outpace existing power sources in highly specialized and extreme environments: in satellites, spacecraft, and Antarctic and lunar-based observation platforms. Small, mobile nuclear reactors used to power batteries and electrical auxiliary systems were developed through the AEC's Systems for Nuclear Auxiliary Power (SNAP) program starting in 1960. Overseen mostly by the Air Force (and later, NASA), the SNAP program worked to miniaturize nuclear reactors for transport on a satellite and later on a manned space voyage.[32] Ultimately, however, the SNAP contractors could not reduce the size and weight of the reactors to the required specification of the Air Force and of NASA, and in 1970, the AEC admitted that SNAP applications within the Apollo lunar program were probably "optimistic by at least five years." The Apollo program resorted to solar panels to generate auxiliary electricity.[33]

The SNAP program produced mixed success, including producing workable reactors for marine buoys, communications satellites, lunar probes, and the Pioneer and Viking spacecraft. The use of reactors in unmanned spacecraft and satellites obviated the need for heavy lead shielding to protect human occupants, but the material challenges of heat transfer out of a reactor core and the sheer weight of the uranium fuel impelled NASA to invest in more advanced solar-voltaic cell technology. The Army's own effort at miniaturized mobile reactors, conducted under the Army Reactor Program, was geared not for space but rather for electric power generation in isolated areas such as at Ft. Greely (Alaska), Camp Century (Greenland), and McMurdo Station (Antarctica). These reactors, which all began operation in 1962, worked well but were ultimately discontinued because of the costs and complications of servicing them.[34] In both the SNAP and Army Reactor Programs, both the strengths and weaknesses of reactor technology were exposed. Reactors were robust and durable. They operated reliably in extremely remote areas such as outer space and Antarctica and could generate electric power for many years without need for fueling or service.[35] But they were expensive to build, could not be shrunk below a certain threshold, and were nearly impossible to refuel once installed far away. That is, the operational cost of reactors was low, but the construction and service costs (both financial and logistical) were high. They were competitive in certain situations where cost concerns were subordinated to longevity but generally lost out to more conventional power sources (such as solar-voltaic cells) once the lifetime operational cost was taken into account.

The Defense Department and the AEC considered the use of nuclear reactors in jet engines starting in 1961. The technology was fairly clear – using an onboard nuclear reactor to superheat air and expel it forcefully out of a nozzle. A plane equipped with the technology would be able to fly without limit with no need to refuel but at the cost of being made to carry both the heavy reactor core and the lead shielding to protect passengers and crew. From the onset, however, military planners were skeptical about the needs of such a technology, confident in the knowledge that aviation fuel could be easily transported to fuel depots around the world or carried on ship for carrier-launched planes. In many ways, the nuclear-powered jet engine appeared to be a solution in search of a problem, and no such problem could be found. No nuclear-powered jet aircraft would ever fly.[36]

The idea of a nuclear-powered rocket was more compelling, given the impossibility of refueling in outer space and the limitations on space flight forced by the capacity of a rocket's fuel reserves. Any rocket needed to expel matter out of its nozzle at tremendous velocity to produce thrust. In a traditional rocket, the fuel served two purposes: as a source of chemical energy from which to derive heat (and ultimately kinetic force) and as a source of mass to produce thrust. In a nuclear-powered rocket the two functions could be separated, with the reactor providing energy to heat up a propellant (most likely liquid hydrogen) and expel it. Such a rocket could, in theory, produce three times the thrust per mass of propellant, but it would need to use at least some of that thrust to counter the weight of the onboard reactor and core.

In the mid-1950s, NASA engineers were enthusiastic about their Project Rover program to develop a nuclear thermal rocket and went as far as testing a series of small reactors (the Kiwi, Phoebus, and Peewee programs) through 1972.[37] The program was regularly slated for termination but earned the strong support of New Mexico senator Clinton Anderson, who represented Los Alamos (where many of the tests were being carried out). In the end, the program encountered many of the same challenges which thwarted nuclear-powered merchant ships. While technologically feasible, a nuclear-powered rocket was so expensive and cumbersome (the rocket needed to be enormous to counter the fixed weight of the reactor) and produced such logistical complications (handling and disposing of radioactive fuel and waste) that NASA dispensed with it. No nuclear-powered rocket ever flew and in 1973, after the last Apollo mission, Congress terminated the program's funding. In fact, analysts had predicted the program's eventual failure as early as 1961, when one internal report estimated the lifetime design and production budget for the rocket at $100 *billion*,[38] while another predicted that enhancements in the design of chemical rockets would probably adequately serve the US rocket program for the foreseeable future.[39]

Other ideas for nonmilitary uses of nuclear power were considered through the 1960s – notably nuclear power for desalination. Although large parts of the United States were well served with freshwater supplies, certain areas (such as the Rocky Mountain west and the Great Plains) existed in semiarid zones where water was scarce. (The Ogallala Aquifer, which was the primary source of both drinking and agricultural water in the Great Plains states, could not be replaced once depleted, and the water of the Colorado River had long since been allocated in full.) Desalination technology existed, but it drew huge amounts of electricity and produced fresh water at a final cost of about approximately $1 per 1,000 gallons. In 1961, the Office of Saline Water in the Interior Department constructed a nuclear-powered plant in Point Loma, California, which produced one million gallons of fresh water daily. This plant was transferred to the Navy in 1963 for use at Guantanamo Bay, Cuba. In 1965, Governor Nelson Rockefeller of New York sought permission from the Department of Interior to develop a nuclear-powered desalination facility on Long Island (which relied on limited groundwater and whose population was expanding rapidly) using a model similar to the Point Loma plant, but this time the Department of Interior refused his request.

Again, the technology was feasible, but the total financial cost was high. A typical conventionally powered desalination plant produced about two million gallons of water per day. A large nuclear-powered desalination plant which could produce 50 million gallons per day *might* justify its construction and servicing costs. One White House report estimated that water coming from such a plant could cost as little as 22 cents per 1,000 gallons. Glenn Seaborg, the chairman of the AEC, was skeptical, noting that the project might work but would be expensive, complicated, and politically vulnerable. "We can't say that there would be *no* useful technological information acquired," he wrote, but that certainly "very little technological information" would be acquired.[40] President Johnson eventually vetoed the project. "Proceed with the plan of having the project reviewed with full explanation for disapproval," he commanded his staff.[41]

Efforts at peacetime uses of nuclear power were only partially successful through the 1960s. While civilian power production was successful, and would continue to be until thwarted by public opposition following the spectacular near meltdown of the Three Mile Island nuclear plant near Harrisburg, Pennsylvania in 1979, other efforts were defeated by basic limitations of reactor physics. Reactors were heavy, required thick shielding, and were expensive to service, clean, and refuel. They produced dangerous radioactive detritus, which needed to be handled and disposed of by highly trained teams using sophisticated equipment. They were challenging to build and install. They worked best when priority was given to optimizing power generation and minimizing refueling – that is, for an enormous naval ship or submarine for which extensive range and high speed was paramount. In most civilian applications, the costs simply did not justify the benefits. As known fossil fuel deposits grew during the postwar decades, and as diesel and gas engines grew more efficient, the competitive advantage of nuclear power diminished. It would find its greatest use in electric power generation for those countries lacking domestic fossil fuel deposits and which did not wish to rely on external suppliers (notably France). But even nations which required substantial imports of fossil fuels often balked at the hazards of operating nuclear power plants and of disposing waste. Japan, a highly industrialized nation with almost no domestic supplies of coal or oil, declared that it would eventually shutter its entire nuclear power industry following an explosion at its Fukushima reactor in 2011.[42]

The Eisenhower administration had prioritized peaceful atoms, but atoms were best used to create explosions and run large power plants. Smaller power plants were expensive with limited applications. But in a series of almost bizarre proposals through the 1950s, the Eisenhower administration approved a series of peacetime applications of nuclear bombs known as Projects Chariot and Gasbuggy.

Plowshare, Chariot, and Gasbuggy

In June 1957, the AEC established a program called Plowshare to investigate the peaceful uses of atomic bombs in excavation. The project would be headquartered at the Lawrence Berkeley Radiation Laboratory and would consider such strange ventures as using atomic bombs to dig harbors and canals and excavate natural gas fields. The proposals were euphemistically termed "geographical engineering" and were confined to the United States with an eye toward economic returns.[43] None of the projects came to fruition, although the United States did conduct a number of retroactive studies on the debris and destruction resulting from underground atomic bomb tests. This era of postwar science can be somewhat risible, but it should be seen as an earnest effort to exploit a powerful new weapons technology toward peacetime uses.

The first of Plowshare's efforts was to excavate an artificial harbor near Cape Thompson in northwest Alaska. The project, named "Chariot", was potentially commercially justified. Substantial oil deposits on the North Slope of Alaska made a viable Alaskan seaport in that part of the state economically attractive, and the proposed site was far from most human habitation. The nearest Inuit (Eskimo) villages were Point Hope and Kivalina, 35 and 40 miles to the north and south respectively. The coast was ice-bound for much of the year, and there were no natural harbors, shipping, industry, or roads. The only contact with the area was through dogsleds and seaplanes.

Chariot, as originally proposed, would explode a 2.4 megaton atomic bomb several miles inland to excavate a harbor and then explode a series of four smaller atomic bombs in a linear pattern to excavate a channel to the sea. In 1958, the AEC requested that 1,600 square miles of land and water in the vicinity be withdrawn from the public domain. Edward Teller, the director of Lawrence Laboratory, led a group of scientists to the area later that year to raise support and enthusiasm and persuade Alaskan residents that the project would generate economic benefits for the state. Business leaders, however, were skeptical of the commercial need for the project (irrespective of other concerns). Senator Robert Bartlett (D., AK) challenged the premise of the project, stating that

> no one on the Commission staff concerned with this believed for a minute private capital would, in the foreseeable future, invest money in the area merely because an artificial harbor had been created. . . . For one, I hope the AEC does its blasting elsewhere.[44]

The AEC downscaled the project the following year so that the largest of the bombs would only be 280 kilotons – still about 20 times as powerful as the Hiroshima blast.

The major problem with Chariot was the danger of fallout deposits. While winds in the area generally blew away from the Inuit villages, and while much of the radiation would be initially contained by the debris falling in on the underground blast craters, substantial radioactive material would eventually escape into the surrounding ecosystem. This posed a problem insofar as the principal flora of the area, lichens, absorbed elemental nutrition directly from the air.[45] Lichens tended to concentrate any radioactivity present in the atmosphere, particularly in the form of strontium-90 (Sr^{90}). The local large fauna, caribou, relied on lichens for a substantial portion of their diet, and the strontium thus became concentrated in caribou tissue. The local Inuit, in turn, relied on caribou meat for their diet, making them vulnerable to radioactive buildup in their tissue.

Tests conducted by the AEC's Division of Biology and Medicine in 1960 found substantially higher levels of Sr^{90} in Alaskan caribou than in domesticated cattle in the lower 48 states. It concluded that the Sr^{90} levels in caribou and reindeer meat were "considerably higher than that of any other animal used for food anywhere in the world," largely a product of fallout from Soviet atomic tests in eastern Siberia that had blown across the Bering Strait.[46] Fallout from the Chariot blasts would almost certainly increase local levels of strontium, and nobody could reliably estimate what the levels would increase to. One scientist investigating the issue suggested that the AEC estimates for strontium fallout from Project Chariot might be off by a factor of ten.[47] An internal AEC report from 1961 admitted that, "because the probable physiological effects of the Eskimo's unique diet caution against direct transfer of conclusions based on Sr^{90} studies of inhabitants of temperate zones, predictions about Sr^{90} absorption by Eskimos cannot now be made."[48] Two outside visitors to Point Hope, James Haddock and Vincent Foster, took interest in the plight of the local Inuit and wrote a persuasive plea to President Kennedy citing radiation poisoning of the lichen, caribou, and marine life, and the fact that the AEC was basing its own radiation estimates on previous underground blasts that were only one-hundredth as powerful. The visitors wrote:

> They [previous blasts] were detonated either in alluvial fill or in bedded turf, not, as here, in perma-frost bordered by pack-ice; they involved

no adjacent ocean supporting a vital food source of a local population, no grazing range of animals on which a local population fed, and no nearby water supply of a local population. Plainly no extrapolation from such experiments can be relied on to predict with certainty that Project Chariot will have no injurious effect on Point Hope or on its food supply or on its water supply.[49]

Project Chariot was cancelled in 1962 for several reasons, most pressingly for its lack of commercial potential. Even in the absence of the project, however, the local ecosystem became contaminated. In August of that year, the AEC transported radioactive material from the Nevada test site to the Chariot site in an effort to study the dispersal of radioactive contaminants on the local environment, leaving a radioactive site which would take a quarter century to remediate.[50] Teller, always bullish on nuclear weaponry of all types, expressed regret at the failure of the project, which he felt would have successfully demonstrated a new use for atomic explosions. In Anchorage, on his tour to promote Project Chariot, he told an assembled crowd, "If your mountain is not in the right place, just drop us a card."[51]

Other uses of atomic explosions were also considered. In Nevada, the AEC conducted Project Cabriolet with a 2.5 kiloton underground detonation to test the cratering performance of underground explosions. Projects Sedan and Danny Boy involved similar testing but in different soils. Project Hardhat used a 4.5 kiloton bomb to create a 63-foot cavern 950 feet below the Nevada desert. Hardhat was deemed a success insofar as it trapped nearly all radiation in the glassified material which formed underground in the heat of the explosion, which could then be removed through conventional mining techniques and stored.[52] Project Gasbuggy, conceived in 1966, was aimed at liberating underground gas reserves in a depleted gas field in Farmington, New Mexico, while Project Buggy involved five simultaneous underground nuclear blasts to excavate a cavern 70 feet deep, 200 feet wide, and 820 feet long.[53] Most ambitious was a plan to excavate a new Panama Canal using nuclear warheads. The plan might have saved two-thirds of the cost of digging a new canal through conventional means, but it was dismissed before it could be seriously considered.[54]

Virtually, no commercial use of atomic explosives emerged from this decade of effort. Although over the course of Plowshare the United States detonated 31 nuclear warheads in 27 different tests, the program never garnered public support and was discontinued in the mid-1970s. All of the Plowshare tests were done in isolated locales, reflecting the inherent danger of radiation release and long-term contamination. As always, fear of falling behind was at least partially responsible for the initiation and survival of the project. Concerns about Soviet scientific and technological prowess drove the United States into projects that had little chance of commercial success, which in turn invited public skepticism and even litigation. Inuit communities were still negotiating with the state of Alaska in the late 1980s for reparations due from radiation poisoning, and the United States issued formal apologies to many Americans who had lived downwind from test sites in Nevada and in the Pacific Islands in 1995. For all the essential nuttiness of the enterprise, though, the Canal Study Commission noted in 1966 that if the United States did not blast a new canal through the Panama isthmus using atomic bombs, someone else would. The chairman of the commission wrote, "The promise of economical large-scale earth moving with nuclear explosives is so great that further development of the technology by some nations is inevitable."[55]

Plowshare did not achieve its stated goals, but it did succeed in drawing the attention of both Congress and the White House to an expanded role for the AEC in a post–Cold War era. The NSF, still very much subordinate to the Navy and the AEC in funding big scientific projects, tried its own hand with Project Mohole – a 1963 effort to drill a hole through the Earth's crust into the semi-molten mantle beneath. The Earth's crust was generally thought to extend between 20 and 25 miles down – far beyond the 1.5-mile range of existing drilling equipment. But beneath deep parts of the ocean, at depths of 10,000–18,000 feet, it was thought that the crust might be as thin as three miles. In 1957, two geologists, Harry Hess of Princeton and Walter Munk of the University of California, suggested to a NSF Earth Sciences panel that punching a hole into the mantle might be the sort of large, era-defining project to draw attention and prestige to the NSF. The project gained support when Hess learned from a Russian colleague that the Soviets already possessed drilling equipment capable of the undersea drilling effort.

Using a small pilot grant from the NSF, the researchers carried out an initial test drill off the coast of San Diego using a mobile drilling ship leased from a consortium of oil companies. With an additional $1.5 million from the NSF, researchers modified the ship to make it more stable in the open ocean and successfully drilled in 10,000 feet of water to a depth of 600 feet.[56] But the next step, drilling to a depth 20–30 times as great, required new equipment and a new contractor. The NSF hired the Houston-based oil field services company Brown & Root and went before Congress to seek additional funding. The House Appropriations Committee agreed to place an additional $5 million into the NSF budget for the following year to fund the project.[57]

The project's budget grew rapidly; within a year Brown & Root raised its estimate for the drilling costs from $35 million to $68 million and within two years to $125 million. In an effort to preserve the NSF budget, Congress created a separate Mohole fund. As costs grew, so did awareness of the technical challenges. In 1966, Congress cancelled the program, having spent $57 million in total on the project.

Big peacetime scientific efforts, divorced from military consideration, were part of a gentle shift in building support for science independent of Cold War concerns. Bush's 1945 vision for the NSF had still not been realized by 1965. In that year, the foundation's budget still totaled only $250 million – a pittance compared to the combined budgets of the AEC, ONR, and the burgeoning NASA. Efforts to convert military research to civilian standing, or to at least find civilian applications for military hardware, largely came to naught. Congress had more than enough political support to continue to fund military research at a high level, and voters' support for military funding had yet to be shaken by the expanding conflict in Vietnam. Plowshare, Gasbuggy, Chariot, and Mohole were diversions from the central task of government science in the 1950s and into the 1960s.

International Cooperation

Nonmilitary scientific efforts during the 1950s and 1960s invited partnering across borders. While the American and Soviet scientific enterprises far outpaced those in western Europe and Japan in the immediate postwar years, those regions also had long traditions of scientific engagement and excellence. Germany and France had reigned supreme in scientific prowess through the 19th and early 20th centuries, and in 1955, western Europe hosted 175,000 scientists and 250,000 engineers – more than the United States or the Soviet Union. What the region lacked, however, was steady scientific funding. While western European governments were able to divert some funds to university-based scientists, the sums paled relative to those invested by the United States, even on a per capita basis. Only Great Britain was investing significant amounts of government funds into research, and here again, it trailed US investment by an order of magnitude.[58]

President Eisenhower's science advisors, working in conjunction with the NSF, recommended in 1955 a series of foreign exchanges and cooperative programs to try to engage European and Japanese scientists in American projects. Such efforts might include visiting appointments, international scientific conferences (both home and abroad), and even American funding of select, nonmilitary foreign research programs. These vague recommendations were clarified over the following years to include cooperation in space science, semester or year-long postdoctoral and faculty exchanges, international cooperation on developing educational material, and international seismological observations stations.[59] The working group also explicitly recommended suspending the visa restrictions which might be preventing such visits and conferences from taking place.[60]

As scientific advisory staff studied the issue, however, their recommendations changed. Simple exchanges and agreements of cooperation would be inadequate given that the primary need of foreign scientists was sustained financial support. Switzerland, for example, "had simply no money at all for research," wrote PSAC member Lloyd Berkner; virtually all of its funding was coming from American grants on military projects.[61] In November 1959, PSAC recommended that the United States commence a broad program of international scientific funding.[62] Such funding could enhance alliances while enriching scientific efforts both at home and abroad. But underpinning the move toward internationalism was, as always, concern over impending Soviet threats. PSAC noted:

> Further, in these days of technological warfare, the strength of our military alliances necessarily depends on adequate research and resulting scientific competence in our allied nations. When support by the United States government is necessary to insure such research, funds should be provided lest our common defense be weakened.[63]

Such efforts were largely focused on wealthy, industrialized nations, but PSAC gave some consideration to supporting scientific education and work in less industrialized regions, particularly in South America. As part of the Act of Bogota of 1961 – a $500 million American commitment to the region for economic and educational aid – President Kennedy included funds for "expanding the wonders of science," educating the region's youth, training science teachers, and laying the basis for future scientific investigations. The President wrote:

> I invite Latin American scientists to work with us in new projects in fields such as medicine and agriculture, physics and astronomy and desalination – to help plan for regional research laboratories . . . and to strengthen cooperation between American universities and laboratories.[64]

The effort was designed to forge stronger bonds with western hemispheric neighbors while also promoting economic and industrial growth. Eugene Skolnikoff, the President's deputy science advisor, justified the support with "the simple fact that fundamental investigation is part of the full intellectual apparatus of any country which hopes to understand and be a part of modern society" and that graduate education was one of the prime means of "ensuring that the small segment of the work force represented by highly trained professional people will be flexible and adaptable to meet changing requirements."[65]

Latin America trailed Europe both in research funding and in university infrastructure. Educational weakness permeated teaching systems across the continent, although wealthier countries such as Venezuela, Argentina, and Mexico claimed a few excellent colleges. Few students graduated high school and few high school graduates attended college as full-time students. Universities lacked adequate library and laboratory resources. Professors frequently moonlighted to supplement their inadequate salaries. Most troubling was the absence of a culture of economic rigor. Skolnikoff identified the "lack of earnest and total dedication by university students to a complete and systematic regimen of studies" as one of the major barriers to continent-wide scientific prowess.[66] PSAC recommended a series of research grants and graduate support fellowships as first steps in strengthening science in the region.

American scientific aid to less industrialized nations was part of a broader commitment to economic and technological development of potential allies. As allies increased in wealth and education, they purchased more American goods, strengthened their armies, and committed themselves to military alliances. They were more likely to align themselves with Western bloc nations and to embrace capitalism as an economic system. Albert Noyes, a chemist and member of the National Academy of Sciences, opined on a proposed technical assistance program to his colleague, Detlev Bronk. He enumerated numerous arguments for American support of science and education in developing countries, including:

> the retention of old friends and the acquisition of new ones . . . the creation of new markets by educating other peoples to desire more food, better clothing and more manufactured goods . . . the maintenance of foreign trade and hence the domestic economy of the United States . . . to develop the natural resources of other countries with the objective of providing new courses of raw materials as our own raw material become depleted or becomes insufficient in the event of war . . . as part of the Cold War in an attempt to convert uncommitted nations to free enterprise and to the ideals of personal freedom and initiative in which we firmly believe.[67]

Much trickier was American scientific cooperation with the Soviet Union itself. Here, the goals were not building an alliance, or even easing tensions in a hostile relationship, but merely maintaining lines of communication insofar as was possible while both nations maintained near states of war. Cooperation here could not undermine national security, jeopardize state secrets, or facilitate scientific espionage. At the same time, the United States prioritized maintaining scientific dominance, leading to a compromised posture on cooperation. One NSF analyst tersely wrote, "Obviously the United States would not support the build-up of the scientific potential of the USSR on any occasion in the foreseeable future."[68] PSAC concurred and noted that the sorts of restrictions which would need to be placed on visiting Soviet scientists might produce the opposite of the intended effect – alienating the visiting scientists through secrecy and limits on travel.

How much could be gained substantively through such an exchange? The Soviets hosted a formidable scientific enterprise as evidenced by their success constructing nuclear arms and major engineering projects. But because Soviet scientists published in their own journals (in Russian), eschewed international conferences (except within the Eastern bloc), and conducted their work under conditions of state security, American scientists simply did not know how much they could learn from their Soviet colleagues. PSAC asserted in the late 1950s that the United States had much to learn from increased contact with Eastern bloc scientists and "much to gain from increased scientific contact." But would the exchanges even lead to sharing of material of scientific value?

That is, would the visiting Soviet scientists have license to share their most interesting results and their most innovative research?[69] In fact, if the goal of the exchange was not so much substantive science but rather enhancing a "spirit of free inquiry," promoting "objective observation," and developing "lasting personal relationships" (in PSAC's assessment), then the veil of secrecy and security which would envelope scientific exchanges in both directions would undermine the program's objective. One PSAC analyst wrote,

> If the visiting Russian finds in the U.S. restrictions patterned after those he suffers at home, and the same tendencies towards political involve- ment of non-political matters, he may not carry back with him the very message we are attempting to get across.[70]

Exchanges with the Soviets were potentially more costly than beneficial as they could undermine existing scientific relationships between the United States and western European nations and Japan. Increased cooperation with the Soviet Union might, in fact, be illusory, while actually working against the scientific boycott which the United States encouraged its allies to maintain with the Soviet Union. Skolnikoff advocated for including Soviet scientists in multinational coalitions and conferences rather than in one-to-one privileged exchanges. Such exchanges could not be justified, Skolnikoff wrote, "in view of our attempt to discourage other nations from having relations with the USSR," and the United States, rather, should focus on big theoretical sci- entific exchanges around particle physics, space science, rocket technology, astronomical observations, and geophysical explorations in the Arctic.[71]

In 1960, the USSR Academy of Science and the [American] National Academy of Science established an exchange program wherein 20–25 sci- entists from each country traveled abroad each year to work in labs of the host country, travel, and meet prominent researchers. Most of the cost of the American side of the program was funded by the NSF, which included housing and meal stipends for the visiting Soviet scientists, some lost salary funds, and even support for accompanying family members. The program was coordinated and monitored by the Soviet Office of the Foreign Secre- tary and by the US Department of State. The two nations were careful to maintain parity in the number of scientists visiting, time spent in specific labs, and invitations from prominent scientists in each country.

Paul Doty, a professor of biochemistry at Harvard, argued that the exchanges were not particularly productive as they paired subpar Soviet sci- entists with innovative American ones. In his view, fruitful and stimulating scientific discussions were possible only between scientists at the same level of talent and accomplishment, and only about one-fourth of scientists in the USSR met this criteria. He wrote,

> To complete the picture, I would judge that nearly half of the remain- der of scientific research in the Soviet Union is carried out in fields

that seem of little importance to us and hence offer no basis for useful inquiry and discussion. The level of research in the other half is generally behind our own efforts and hence not particularly interesting as far as basic science is concerned.[72]

While many participants in the exchanges found them to be worthwhile, they tended to work best when the American scientists were internationally renowned and thus given special treatment by Soviet hosts and granted better access to Soviet labs. In contrast, most visiting American scientists were routinely denied access to the labs where the most promising and innovative science was transpiring. Doty judged that the visits produced a "low level of critical discourse, indicated by the paucity of meetings, the absence of migration of scientists from one institute to another, and the lack of reviews and analyses of their own work."[73] While the visits were not useless, they were hardly ideal; linguistic barriers, security concerns, and general opacity limited the benefit of the program. Doty concluded, "Except for the small working conference attended by an American with high fluency in Russian, it is unlikely that attendance at Soviet meetings would be worth the time."[74]

More successful were the American cooperative agreements with western European nations to develop and fuel nuclear power plants. In 1958, the United States committed to working with France, West Germany, Italy, and the Benelux countries – the six member states of the European Atomic Energy Community (EURATOM) – to help them build six nuclear power plants, by 1963, which would provide electric power for five million residents. High costs of electric power in Europe justified the $350 million construction cost of the plants which were funded, in part, by a low-interest loan from the United States. The plants would also become customers for American-enriched fuel, of which the United States would sell 30,000 kilograms over 20 years. The United States would also take back the spent fuel to either repurpose or dispose it.[75]

European construction of power plants allowed European nuclear engineers to produce innovations in reactor design. By 1964, Spain contracted with France (rather than the United States) to build that nation's first reactor in Cataluna, from which France would buy back about half of the generated power. That same year, Canada began supplying technical aid to Pakistan for its own nuclear power effort, and Israel began negotiating with the United States for support in building a nuclear-powered desalination plant.[76] At the same time, France was preparing to install mobile reactors into three nuclear-powered submarines.[77] By 1966, and possibly before, European reactor technology surpassed American technology, particularly in the area of breeder reactors, which Europeans saw as the solution to the uranium monopoly which the United States continued to hold.[78]

Foreign reactors were steady customers for American fuel. In 1956, the United States initiated a program to sell 20,000 kilograms of enriched fuel to EURATOM members over the next decade to power reactors which were then in the planning stages.[79] Two years later, this program was expanded to include 1,000 kilograms of plutonium shipments for use in "fast" reactors, and in 1966, the program was expanded again to include an additional 500 kilograms of plutonium.[80] Other programmatic expansions included uranium distribution to Latin American countries (notably Argentina) starting in 1962.[81] The program rested on commitments of Western allied nations to the principles of nonproliferation, inspection by the IAEA, and an unspoken agreement to maintain US monopoly on fuel enrichment. The agreement would end in 1974, when French President Pierre Messmer committed to a fully nuclear French electric power industry in the wake of steep rises in oil prices.

New Accelerators

"Atoms for peace" was most fully realized in the nation's new civilian power sector, but continued work with subatomic particles was part of the effort. The NSF continued to fund basic science in the physical sciences during this time and paid particular attention to both the largest features of the universe (stars and galaxies) and the smallest (muons, pions, and neutrinos). Paradoxically, both areas of focus required enormously expensive pieces of equipment: optical and radio telescopes in the first case and particle accelerators in the second.

Each new stage of particle research required more powerful accelerators to smash atoms to ever smaller pieces from which to gain new understanding of the basic state of matter. From Lawrence's first million-volt cyclotron, which he built at Berkeley in 1931, postwar physicists built new machines, which could raise energy levels first to hundreds of millions of electron volts and then to billions of electron volts (BeV). By 1965, the Japanese had completed a 40 BeV machine, and the Soviets were nearing completion on a 70 BeV proton machine. Americans and Europeans (through CERN) envisioned machines in the 200–300 BeV range in the near future.

Research in the physical sciences was limited more by fiscal constraints than intellectual ones. Time on the nation's large accelerators needed to be reserved months in advance, and many particle physicists were forced onto large teams lest they be denied access to the machines. A special planning committee, led by Emanuel Piore, the director of research at IBM, recommended in 1958 that the United States begin planning for more substantial accelerators to meet the needs of physicists in the late 1960s. Donald Hornig, who succeeded Jerome Wiesner as the science advisor to President Kennedy, testified to Congress that higher-energy accelerators were the single most important step for the government to undertake to advance physics.[82]

Large particle accelerators were gigantic government-funded construction projects which created design and construction jobs initially, and then jobs to support the community of researchers who were drawn to work near the new facility. The most glaring need for a new accelerator was in the upper Midwest, where many of the nation's most formidable public research universities were located.[83] The region's research universities coalesced into the Midwestern University Research Association (MURA) to lobby aggressively for the proposed 200 BeV reactor recommended by the Piore Committee. MURA argued to the Kennedy White House that while the East had four major accelerators with two under construction, and while the West had two accelerators with two more under construction, the Midwest had none. Besides disadvantaging Midwestern universities, the absence of an accelerator also undermined the region's ability to attract research-based companies which tended to situate themselves near large pools of research talent. A group of Midwestern Senators lobbied hard for the new accelerator, suggesting that the region's economic future was tied to the proposal.[84]

The Kennedy administration initially agreed to the proposal and selected Weston, Illinois, about 30 miles west of Chicago, as the site for the accelerator.[85] The new machine would be the nation's largest piece of scientific apparatus, nearly one mile in diameter built on 6,800 acres. In 1967, however, President Johnson terminated the effort, feeling that the cost of the project could not justify the investment in the midst of civil rights protests and deeper military involvement in Vietnam. While the region's universities could draw on the much smaller existing accelerator at the Argonne National Laboratory (in Lemont, Illinois), schools in the area would have to continue to reserve time on the coastal machines, weakening the region's stature in particle physics.[86]

The loss of the Midwest accelerator was particularly acute when juxtaposed against the Stanford Linear Accelerator (SLAC) completed in 1967. SLAC, at two miles, was the world's longest linear accelerator and promised to make the Bay Area a continued destination for the nation's best researchers in particle physics. Moreover, it reinforced Stanford's position as one of the nation's preeminent research universities, adding basic research to its already formidable engineering work and pulling it into the front ranks of particle physics. Spill-off from SLAC might be expected to add to the region's reputation as a destination for the most innovative technology companies which could employ the nation's brightest young physicists and electrical engineers. Hornig noted that SLAC allowed Stanford to perform as a "university in a national framework" and singled out the user groups which were already migrating to the area.

The government's decision to construct the Stanford accelerator while cancelling the Midwest accelerator exacerbated the growing divide between the nation's elite research universities and all others. Even as Hornig noted the need of advanced students to "learn by doing" in his SLAC dedication speech, he ignored the fact that Stanford students were already able to do this better than most, given the remarkable resources that the university was able to place at its students' disposal. SLAC was a new chapter in an evolving story of rich and powerful research universities getting richer and more powerful through the steady support of government scientific largesse. Government resources continued to flow to the "most creative, energetic, and effective universities," even while pedagogy demanded that advanced undergraduate and graduate students across the country have access to research-active faculty and well-equipped laboratories.[87]

While military applications continued to drive the preponderance of government scientific largesse through the 1950s and into the 1960s, peaceful considerations were not trivial. The long-term promise of nuclear-generated electric power, the burgeoning role of the United States as supplier of radio-isotopes and enriched atomic fuel, and the continued support of theoretical research and civilian engineering applications were important components of America's work in the physical sciences. Planners questioned, however, if the government could politically afford to sustain major commitments to research without clear application. The failure of the 200 BeV accelerator at Madison was a case in point. If only military, or quasi-military, projects could garner political support and sustained funding, then military needs would remain paramount in driving American physical science in coming decades.

Notes

1 Robert McKinney, "Panel on the Prospects for Civilian Atomic Power," DDE, PSAC, 4: Nuclear-Plowshares. Also, McKinney to Killian, January 9, 1959, found in the same folder.
2 The group of six, known as "the Martians," included Wigner, John von Neumann, Leo Szilard, Edward Teller, Paul Erdos, and John Kemeny. All had emigrated to the United States in the 1920s and 1930s, largely to escape fascist anti-Semitic persecution.
3 Eugene Wigner, "Statement Before Joint Committee on Atomic Energy," *IIR*, 17:8, July 9, 1953, p. 5.
4 "What Is the Atom's Industrial Future?" *Business Week*, March 8, 1947. Also, "The Atomic Power Program," *New York Times*, February 9, 1962.
5 "Nuclear Electric Power Program," March 4, 1961, JFK, WHCF, 10: AE.
6 Quoted in an internal White House memo to McGeorge Bundy, May 3, 1963, JFK, WHCF, 10: AE.
7 "The Atom's Industrial Sponsors," *Business Week*, March 22, 1947.
8 Robert Wilson, "Expected Impact of Nuclear Energy on Our Economy," April 19, 1963, LH, 21: "Speeches by Others."
9 "NJ A-Plant Promises 4-Mill Power," *APPA*, appended to White House memo of June 12, 1964, LBJ, WHCF, AE, 2: AT2. Also, Cater to Owen, May 25, 1964, LBJ, WHCF, AE, 2: AT2.
10 International Atomic Energy Agency, "The Jersey Central Report: New Claims for Competitive Nuclear Power Raise Questions," *Atomic Power Newsletter*, March 31, 1964.
11 Johnson to Seaborg, April 17, 1965, LBJ, WHCF, FG, 261: AEC, p. 2.
12 Seaborg to Johnson, October 24, 1967, LBJ, CF, 109: AEC 1967.
13 Seaborg to Johnson, January 9, 1968, LBJ, CF, 109: AEC 1967.
14 "Atomic Energy Commission and NASA," March 4, 1961, JFK, WHCF, 10: AE.
15 Kennedy to Holifield, no date (1961), JFK, WHCF, 10: AE.

16 Seaborg to Johnson, May 12, 1964, LBJ, CF, 108: AEC 1963–1966. Of note, in 1965, President Johnson wrote to Seaborg extolling the "development of the advanced converter and breeder reactors . . . required for the more efficient and economical use of our Nation's nuclear fuel resources." This may have simply been a misunderstanding on Johnson's part. Johnson to Seaborg, April 17, 1965, LBJ, WHCF, FG, 261: AEC, p. 2.

17 All contracts were contingent on the lessee returning depleted material, including plutonium, to the AEC.

18 Seaborg to Johnson, June 27, 1967, LBJ, CF, 109: AEC 1967.

19 "AEC Price for Selling Uranium Enrichment Services," September 16, 1967, LBJ, WHCF, CF, 31: AEC.

20 Seaborg to Rostow, March 10, 1967, LBJ, NSF, 29: Nuclear Weapons. Also, Kemeny to Rostow, March 24, 1967, LBJ, NSF, 29: Nuclear Weapons.

21 For background on the AEC's radioisotope distribution, see Advisory Committee on Human Radiation Experiments, *Final Report*, GPO, 1994, chapter 6. The committee's discussion of the program focuses on the ethical dimensions of using the radioisotopes in human research but presents a helpful background to the program.

22 War Department, "Announcement of First Shipment of Radioisotopes from Manhattan Project Clinton Laboratories," August 2, 1946, HST, PSF: Atomic Bomb Press Releases.

23 AEC, "AEC Announces Distribution Program for Cyclotron-Produced Radioisotopes," July 24, 1949, DB, FA 965, 28: AEC 1949.

24 *Atomic Energy Newsletter*, March 15, 1949.

25 AEC, "The University and the Atom," October 2, 1950, DB, FA 965, 28: AEC 1950.

26 Gordon Dean, "Remarks Before the American Medical Association," June 26, 1950, HST, SRF, 13.

27 Ibid., pp. 3–4.

28 AEC, "The Impact of Radioisotopes in Cancer Research," December 28, 1951, DB, FA 965, 28: AEC 1951.

29 This example rings of 1960s excessive zeal for modernity. The antinuclear movement which began to question the long-term harm of radiation exposure did not take root until the mid-1970s. It is important to realize, however, that radiation scientists, from early on, understood the potential danger of radiation and radioisotopes, as well as the potential benefit. See Seaborg, "Anniversary of the First Shipment of Radioisotopes from Oak Ridge National Laboratory," LBJ, WHCF, AE, 1:8/12/66.

30 Shields Warren, "Statement to Congress," August 4, 1949, DB, FA 965, 28: AEC 1949, p. 3.

31 The original Savannah was really a hybrid steam-sail design. Her initial crossing, from Savannah, Georgia, to Liverpool, England, took nearly a month. During that time she only moved under steam power for less than four days. The first ship to cross the Atlantic predominantly under steam power was the Dutch-owned *Curacao*, which crossed from Rotterdam to Dutch Guyana in 1827 in 11 days.

32 See AEC, "SNAP: A Report by the Commission," February 1964.

33 AEC, "SNAP: An Evaluation," January 1964, p. 12.

34 Report to the Joint Committee on Atomic Energy, "Army Reactor Program," LBJ, NSF-Charles Johnson, 30: Nuclear – Army Reactor Program.

35 Jackson to Kennedy, February 21, 1961, JFK, WHCF, 10: AE.

36 See "Nuclear Electric Power Program, March 4, 1961, JFK, WHCF, 10: AE.

37 The nuclear-powered rocket programs were grouped under the rubric NERVA (Nuclear Engine Rocket Vehicle Application).

38 This staggeringly high estimate is correct. See "Rover Panel Report," March 21, 1961, JFK, POF, 86: PSAC, 1/61–3/61, p. 4.

39 AEC and NASA, "Project Rover," Arch 4, 1961, JFK, WHCF, 10: AE.
40 Seaborg to White, August 12, 1965, LBJ, WHCF, AE, 2: AT2 Industrial, p. 2.
41 White to Johnson, September 21, 1965, LBJ, WHCF, AE, 2: AT2 Industrial.
42 Hiroko Tabuchi, "Japan Premier Wants Shift Away from Nuclear Power," *New York Times*, July 13, 2011.
43 "Plowshare Program: Fact Sheet on Project Chariot," JFK, WHCF, 10: AE.
44 Quoted in Paul Brooks and Joseph Foote, "The Story of Project Chariot," *Harper's*, April 1962, p. 65.
45 Lichens are not plants; they are a biological community composed of cyanobacteria or algae living symbiotically with fungi, in which the bacteria produce carbohydrates for the fungi to metabolize in exchange for the structural protection granted by the fungi.
46 "Project Chariot," *Nuclear Information*, 3:4–7, June 1961, pp. 3–4.
47 Ibid., p. 4.
48 Ibid., p. 11.
49 Haddock and Foster, "The Threat of Project Chariot to the People of Point Hope, Alaska and Their Way of Life," February 1961, JFK, WHCF, 10: AE, p. 3.
50 The material was removed by the state of Alaska 30 years later. See Dan O'Neill, *The Firecracker Boys*, Basic Books, 1995.
51 Quoted in Brooks and Foote, "The Story of Project Chariot," p. 67.
52 "Plowshare Mining Study in Progress at Nevada Test Site," appended to Clark to Salinger, June 25, 1962, JFK, WHCF, 10: AE.
53 Fouchard to Hackler, March 6, 1968, LBJ, WHCF, CF, 31: AEC.
54 Anderson to Katzenbach, "AEC Plowshare Nuclear Cratering Experiments," October 26, 1966, LBJ, NSF, 29: Plowshare Events.
55 Ibid., p. 4.
56 Daniel Greenberg, "Mohole: The Project That Went Awry," *Science*, 143:3602, January 10, 1964, pp. 115–19.
57 Herbert Solow, "Now NSF Got Lost in Mohole," *Fortune*, May 1963. Brown & Root had powerful political patrons. The chairman of the Appropriations Committee, Albert Thomas, represented Houston in Congress, and President Johnson had long relied on the company for substantial campaign contributions.
58 "Achieving and Maintaining United States and Free-World Technological Superiority Over the USSR," December 3, 1955, MIT, 195: SAC.
59 Skolnikoff to Wiesner, February 3, 1961, DB, FA 965, 16: Panel on International Science.
60 NSF, "Preliminary Report on the Rose of the Federal Government in International Science," December 1955, DB, FA 965, 15: PSAC.
61 Berkner to Skolnikoff, October 19, 1959, DB, FA 965, 16: Panel on International Science, p. 2.
62 PSAC – Science and Foreign Affairs Panel, "Direct Support of Scientific Research in Foreign Countries," November 1959, DB, FA 965, 16: Panel on International Science.
63 PSAC, "Science and Foreign Affairs," *IIR*, 45:10, October 21, 1959, p. 5.
64 Transcribed by Skolnikoff, "Programs of Science and Technology with Latin America," April 22, 1961, DB, FA 965, 16: Panel on International Science, p. 1.
65 Ibid., p. 3.
66 Ibid., p. 6.
67 Noyes to Bronk, July 30, 1958, DB, FA 965, 16: Panel on International Science, p. 1.
68 NSF, "Preliminary Report on the Rose of the Federal Government in International Science," p. 18.
69 PSAC, Science and Foreign Affairs Panel, "Scientific Fellowship and Exchange-of-Persons Programs," November 1959, DB, FA 965, 16: Panel on International Science, p. 2.

70 PSAC, "Reciprocity Policy in Scientific Exchanges with the USSR," November 3, 1959, DB, FA 965, 16: Panel on International Science, p. 4.

71 Eugene Skolnikoff, "International Scientific Cooperation – Discussion Paper," January 11, 1960, DB, FA 965, 16: Panel On International Science, p. 7, 17, 18.

72 Paul Doty, "A Critique of Scientific Exchanges with the Soviet Union and the Possibilities of Changing Their Character," February 10, 1964, DB, FA 965, 16: Panel on International Science, p. 6.

73 Ibid., p. 8.

74 Ibid., p. 9.

75 AEC, "United States-Euratom Program," RC, 7: Press Releases 1956–58.

76 Israel possessed two reactors at the time: a small research reactor near Yavne (often called the "Sorek" reactor for a nearby stream) and a larger reactor in Dimona, which it used to produce plutonium for its nuclear arms program. The desalination project with the Americans may have been a cover for an additional arms-related project. See Seaborg to Johnson, June 9, 1964, LBJ, CF, 108: AEC 1963–66.

77 Seaborg to Johnson, October 13, 1964, LBJ, CF, 108: AEC.

78 See John Finney, "Euratom Seeking Record Purchase of U.S. Plutonium," *New York Times*, May 9, 1966, p. A1.

79 Eisenhower, "Statement by the President," February 22, 1956, RC, 7: Press Releases 1956–58. The program was primarily aimed at the EURATOM members but included the full 37 member countries of the IAEA excluding those capable of producing their own uranium.

80 Seaborg to Johnson, July 13, 1966, LBJ, WHCF, AE.

81 Kennedy to Seaborg, June 21, 1962, JFK, WHCF, 10: AE.

82 Donald Hornig, "Statement Before the Joint Committee on Atomic Energy," March 3, 1964, LBJ, DH, box 7, p. 9.

83 The American south also had no accelerators, but that region of the country had fewer research universities hosting active physics programs. Atlanta made a bid for the accelerator in 1965, having just made substantial investments in its airport and boasting a new professional sports stadium. See "A Proposal for the 200 BeV Proton Accelerator," appended to Patterson to Moyers, June 18, 1965, LBJ WHCF, FG202, 263:6/17/65–7/2/65.

84 Humphrey, Dirksen, et al. to Kennedy, August 6, 1963, JFK, WHCF, 10: AE, p. 2.

85 Bryce Nelson, "200 BeV Accelerator: Moving Into a Wasp's Nest," *Science*, 156:3783, June 30, 1967, pp. 1713–1716

86 "AEC Selects Site for 200-BeV Accelerator," LBJ, WHCF, FG202, 264:5/13/66–12/31/66.

87 Donald Hornig, "Dedication of the Stanford Linear Accelerator," September 9, 1967, LBJ, DH, 8: Addresses and Remarks, p. 9.

9 Satellites, Rockets, and Thinking About Space

Sputnik

On October 4, 1957, the Soviet Union launched Sputnik, the world's first artificial satellite. The 185-pound aluminum sphere, a bit less than 2 feet in diameter, moved over the Earth from pole to pole at a speed of 17,000 miles per hour. At an altitude ranging from 400 to 600 miles, it completed a full orbit every 96 minutes. It carried two radio transmitters which transmitted signals back to the Earth from its most notable features: four 10-foot-long antennae poking from its surface.

Sputnik lacked a camera or transmission relay device, making it useless for either spying or communication. Nonetheless, its launching electrified the American public. Within six months, 91 percent of Americans claimed familiarity with its existence, and half of respondents in a national survey viewed the launch as the beginning of a competition to master space – a competition in which the United States already lagged.[1] But while the American public sensed a grave threat, scientific and political leaders viewed the launch as more symbolically important than tactically so. The disconnect between public perception and expert assessment led to convoluted messages and responses in the first few days after the launch. Henry Smythe, a nuclear physicist, mused to a gathering of his peers that "the shadow of the Russian satellite is surprisingly large and persistent for so small an object," and President Eisenhower's chief of staff, Sherman Adams, remarked to the press that the US space program was aimed at advancing science, "not high score in an outer space basketball game."[2]

Sputnik threatened Americans' sense of security not for what it was but rather for what it represented. The rocket that had launched Sputnik had been powerful enough to lift a substantial object hundreds of miles above the Earth at a speed just short of escape velocity (the speed at which an object could escape the Earth's gravitational pull). The satellite was gathering data about electromagnetic fields in the ionosphere: a subject critical to future development of missile guidance systems. The Soviet government suggested that the nation's scientists were using knowledge gleaned from the satellite to plan manned missions to the Moon and to Mars: to "pave the road to

DOI: 10.4324/9781003363897-9

interplanetary space," in the words of a Soviet press release. And, in a direct shot at American prestige, Soviet scientist K. E. Tsiolkovskii claimed: "Of one thing I am convinced – the Soviet Union will be first."[3]

President Eisenhower initially underestimated the effect of Sputnik on the American public. Assured by his advisors that the American space effort was making steady progress toward its own satellite launch, and comfortable with the rate of technical progress in the nation's rocket development, Eisenhower chose to de-emphasize the notion of a space race and rather to focus on the nonmilitary scientific opportunities for international cooperation and discovery. On November 7, he assured the nation in an Oval Office address that the nation's defenses were adequate and secure and that the Army and Navy had the power to bring "near annihilation to the war-making capabilities of any country."[4] Defense research and development was being funded at over $5 billion each year, with over $1 billion going annually to the development of long-range ballistic missiles. The Navy was developing a system of submarine-launched nuclear-tipped missiles, and the Army and Air Force had hundreds of different short- and medium-range missiles which together contained many times the explosive power of all of the artillery used in World War II.[5] The Air Force had already replaced nearly all its B-36 bombers from the Korean War with longer-range B-52s, which would soon give way to a new series of supersonic B-58s.[6] President Eisenhower asserted,

> In numbers, our stock of nuclear weapons is so large and so rapidly growing that we have been able safely to disperse it to positions assuring its instant availability against attack. . . . Our scientists assure me that we are well ahead of the Soviets in the nuclear field, both in quantity and in quality. We intend to stay ahead.[7]

But such assurances belied the Soviets' many successes. Soviets scientists were building the largest proton accelerator, the most intense neutron source, the largest optical and radio telescopes, and the first long-range ballistic missiles. Lloyd Berkner, a member of PSAC, concluded that while the West still held the scientific crown, the rate of Soviet progress assured its "inevitable rise to pre-eminence" over time.[8] And while Eisenhower claimed that Sputnik had not raised concerns "one iota," most Americans, and many members of Congress, felt differently.[9]

Eisenhower responded by emphasizing the need to invest in American science education and to raise the stature of science in American life. Repeating a trope common since the start of the Cold War, Eisenhower portrayed a nation of youth distracted by proms, pep rallies, football games, and vocational callings, even while its Soviet counterparts focused assiduously on math and hard science. Americans needed better science teaching, better teacher development programs, more generous support of scientific research, and greater respect for the scientific enterprise. While such a turn would take years, Eisenhower agreed to immediately create a new special assistant to the President for science and technology, to enlarge the existing PSAC, and to pull PSAC and the new Special Assistant from the Office of Defense Mobilization into the White House staff. James Killian, the chair of PSAC, would be the first special assistant.

Not surprisingly, missiles and rockets loomed large. If Cold War tensions over the previous decade had played out along the development of nuclear arms, those tensions now pivoted to developing more powerful, reliable, and accurate missiles and rockets. The President directed the Pentagon to appoint a new guided missile director who would answer directly to the Secretary of Defense.[10] The Secretary would mandate new coordination between the incessantly squabbling branches, each of which had been developing its own rockets and missiles and replicating each other's efforts.

Sputnik had the effect of elevating science in the national conversation. The satellite, small though it might be, transgressed international borders and redefined territorial integrity. Soviet scientists had outpaced their American counterparts, creating a sudden impetus at home to invest in science and to celebrate scientists. Thomas Gates, Eisenhower's first Secretary of Defense, recalled the shift in the zeitgeist:

> All of a sudden the scientists became very important. . . . They had great veto power. They became important people. You paid a lot of attention to the extreme people, like Edward Teller, . . . Johnny von Neumann . . . Wernher von Braun, Jim Killian, [George] Kistiakowsky – these people became very important . . . The world really completely changed.[11]

Eisenhower was treading carefully. On the one hand, he cleaved to his initial response of denying the existence of a space race. The United States was working on its own satellite technology and launching vehicles in an effort to expand scientific inquiry into space, and it welcomed international cooperation in this effort. In this sense, Sputnik was not a victory for the Soviet Union but rather a victory for all of humanity. Eisenhower admonished, "What the world needs today, even more than a giant leap into outer space, is a giant step toward peace."[12] On the other hand, the public demanded an immediate American response to the very visible success of Sputnik, and simply committing greater resources to scientific education and reorganizing the executive science advising apparatus was thought by many Americans to be inadequate. The nation would need to put a satellite very quickly into orbit with an American-built rocket, while also demonstrating that American rockets could perform equivalently, or better, than the Soviets' rockets. While the United States had been pursuing both tasks for nearly a decade, Sputnik fundamentally shifted American scientific focus and emphasis while further hardening the relationship between the physical sciences and the military. The Sputnik era would extend the existing model of military funding of university-based research as mediated by industry and the national labs, but the goal of that science would shift from atoms to space.

Thinking About Space

Although an American, Robert Goddard, flew the first successful liquid-fueled rocket in 1926, the modern rocket age can be better traced to the German rocketry program of World War II. The program, based in Peenemünde, Germany, succeeded in building and launching over 3,000 V-2 guided missiles toward London and Antwerp and killing over 9,000 civilians.[13] The V-2 was the first man-made object to achieve supersonic speed, and the first rocket-powered craft which could credibly claim to be guided, controlled as it was by a series of vanes attached to internal gyroscopes. Powered by a mixture of alcohol and liquid oxygen, the rocket accelerated rapidly for its first 65 seconds of flight and thereupon went ballistic, gliding in a long arc toward its target, often at an altitude of 50 miles or higher. Although the weapon did not prove strategically decisive in the war, its success suggested that future iterations of guided missiles might become increasingly important to the military. Notably, one of the limitations of the program was the fact that the alcohol which fueled it was distilled from potatoes: a critical foodstuff for the German population in the waning days of the war. Hitler remarked that had he had the V-2 in his possession in 1939, he could have achieved his objectives without ever launching a war.[14]

German rocket experts who had worked on the V-2 program were recruited to the United States at the end of the war through a program known as Project Paperclip, and assigned to the Army's Redstone Arsenal in Huntsville, Alabama. The group of engineers, led by Wernher von Braun, produced the Redstone missile based largely on V-2 designs. The Redstone, in turn, gave rise to a series of increasingly accurate intermediate-range missiles, which the Army tested through the 1950s. The Navy and the Air Force followed with their own missile programs, producing successively the Viking, Atlas, Thor, and Titan missiles, each with increasingly longer range, greater reliability, and greater accuracy. By the late 1950s, plans were in place to produce long-range missiles which could nearly escape the Earth's gravity and then descend in a long, graceful arc to hit an enemy target a continent away. Such missiles, demarcated as "intercontinental ballistic," became the goal of military hardware designers in the mid-1950s as they were capable of carrying nuclear warheads to targets thousands of miles away while evading antiaircraft fire. If perfected, they could render both Navy and Air Force tactics and weapon systems obsolete.

Missile technology enabled planners to reconsider space travel. The American Rocket Society in 1957 reported that extensive space flight was now "practicable, useful, and economically feasible" and recommended the establishment of a national space flight program dedicated wholly to the new field of astronautics (as opposed to aeronautics). The guiding goal in this venture was a specific velocity – 18,000 miles per hour – the speed necessary for a spacecraft to escape the Earth's gravity and achieve interplanetary travel. The goal was hardly unattainable; recent ballistic missiles had achieved speeds of nearly 90 percent escape velocity in test flights. Goals for the society included thousand-pound launch payloads in five years, hundred-pound orbital payloads in ten, lunar orbits in 15, and interplanetary travel shortly after. Manned spacecraft would follow closely, with a manned lunar orbiting in 15 years and a manned lunar landing in 20.[15] The goals of the program varied widely and included agricultural surveillance, weather forecasting, communications enhancement, research in material science at low temperature and pressure, and biological responses to low gravity. Overriding it all, the society endorsed the "sense of participation in a great adventure" – spaceflight was the new frontier.[16]

In March 1958, Herbert York, James Killian, and Edward Purcell of PSAC reported on space exploration to the National Security Council. While echoing conclusions and recommendations of the Rocket Society, they emphasized that a space program could produce both civilian and military benefits while also promoting basic science. A satellite placed in orbit at an altitude of 22,000 miles, for example, would remain stationary over a single point on the Earth and thus be able to relay radio signals around the curvature of the horizon.[17] Such a satellite could carry a camera and a transmitter and thus transmit pictures of any part of the Earth's surface that lay beneath its orbital path. If high enough, the satellite, or perhaps an orbiting spacecraft, could fly above the Earth's atmosphere and take much clearer pictures of planets and asteroids while also detecting the ultraviolet light permeating intergalactic space, which was normally too scattered to be detectable. Spacecraft flying above the thin layer of ionized matter known as the ionosphere, would be able to detect the streams of ionic particles regularly emitted by the sun. And farther up, at an altitude of perhaps 100 miles, in the vacuum of space itself, a satellite could detect and observe new types of radiation and energy. York and Purcell noted that "Man lives under a permanent 'tent' which is in many ways opaque."[18]

Space flight promised not just upward vision but downward as well. From a vantage point hundreds of miles above the Earth's surface, scientists would be better able to see patterns of clouds, sea ice, and ocean currents. The Earth's planetary systems might become more visually available, leading to unanticipated breakthroughs in geology, meteorology, oceanography, and ecology. PSAC suggested, "This is science as pure as it can be. It will vastly extend the horizon of man's knowledge concerning the universe as a whole."[19]

From orbit flowed travel. Although the moon was biologically and geologically uninteresting, Mars was not. Telescopic analysis indicated the one-time presence of water and a nascent atmosphere. The planet had a similar composition and gravitational pull as the Earth, leading scientists to speculate that it could harbor life. Venus too beckoned, invisible beneath its dense clouds. While temperatures there seemed inhospitable to life, or even simple organic molecules, it was possible that in its sulfuric atmosphere alternative organisms had evolved.

Space flight was predicated on extending missile technology to larger, multistage boosters which could shed the deadweight of spent stages to lift payloads successively higher. For flights to the moon, including a soft landing (which would require yet another rocket to counter lunar gravity), rocket engines would need to become more efficient, fuel tanks much larger, materials stronger and lighter, and the whole system safer and more reliable. Planners at this early stage hypothesized about stacking various Atlas and Titan rockets atop each other, adding supplementary fuel tanks and supercharging the boosters with multiple rocket engines. New fuels based on fluorine rather than hydrogen might produce greater thrust for the same weight while new materials might better contain the intensely cold liquid oxygen needed to combust whatever fuels were eventually chosen. However it was done, the size and cost of a moonshot would be immense. The planning group of 1958 estimated an initial launch weight of 1.5 million pounds and a minimum development time of ten years.[20]

These early plans for space travel dovetailed with the nation's declared International Geophysical Year (IGY), conceived at the Maryland home of James van Allen, a physicist attached to Johns Hopkins's Applied Physics Laboratory. Along with British physicist Sydney Chapman and American physicist Lloyd Berkner, Allen advocated for an international effort to map the Earth's surface using a series of observation satellites launched atop repurposed V-2 rockets. The IGY was eventually elongated to 18 months (July 1957–December 1958) and work begun on a suitable satellite payload and a powerful-enough rocket booster. With the Army advocating for its Redstone, and the Navy answering with a campaign for its Viking, the Naval Research Laboratory submitted a proposal in 1955 for a civilian program of scientific satellite-based exploration, divorced from military control but leveraging the available technology of the Army and Navy missile programs. The White House responded on July 28 with a carefully worded missive of support:

> The President has approved plans by this country for going ahead with the launching of small, Earth-circling satellites as part of the United States participation in the International Geophysical Year. . . . The President expressed personal gratification that the American program will provide scientists of all nations this important and unique opportunity for the advancement of science.[21]

Shortly after, a White House advisory group recommended moving ahead with the Navy's Viking missile proposal, now named Project Vanguard. In his support for the program, President Eisenhower explicitly endorsed the effort as part of a space program which would have both defense and scientific components. "Our planning and our programs must recognize both of these objectives," he stated at an early stage of Project Vanguard. "Our plans for organization will provide for them."[22]

In 1958, Eisenhower recommended that the research functions that had previously been loosely coordinated by the 50-year-old National Advisory Committee on Aeronautics (NACA) be bonded together under a new National Aeronautics and Space Administration (NASA). At the same time, the President created a new Director of Defense Research and Engineering as part of the 1958 Defense Reorganization Act to better unify and coordinate defense-related research across the services. The two actions together clearly separated military-related from civilian-related aerospace research and turned over the Huntsville rocket team to NASA while reserving all missile-delivered weapons systems to the DoD – particularly centered in the new Advanced Research Projects Agency (ARPA).[23] In fact, the two agencies were never quite as independent as intended, with each sharing technological innovations with the other and with the bulk of research and development funding going to ARPA in the early years. Internal White House directives from that time demonstrate the difficulty of cleaving the two efforts cleanly, diverting "non-military space projects such as lunar probes and scientific satellites" from ARPA to NASA, and appropriating "super-thrust" experimental engines from the Air Force for use in the space program.[24] Historian Audra Wolfe summarizes the true mission of NASA, if not the explicitly stated one, as "basic science, prestige, and cover for military missions."[25]

From its inception, NASA maintained traces of its military roots. Although staffed purely with civilians, it retained a hierarchical mindset with a mission-focused culture more typical of the Army than the academy. Its core technology, the rocket thruster, had been developed in wartime Germany and transferred to the United States through a victor's appropriation (of a sort) after World War II. The first decade of its work had been nurtured by the Army's Huntsville operation, and it remained in stiff technological competition with the Navy through its early decades. Its marquis programs – the manned rocket launches of Mercury, Gemini, and Apollo – drew most of their pilots and astronauts from the Navy and Air Force. Although it sought guidance from the (civilian) National Space Board of the National Academy of Sciences, it was very much a creature of Cold War competition with the Soviets: an odd scientific counterthrust to rising Soviet military–scientific supremacy. Lee DuBridge, president of the California Institute of Technology, rued that "prestige and competitive factors" forced NASA to move "too far too fast, and to spend too much money and devote too much effort to the 'spectacular' as contrasted to the purely scientific ventures."[26]

At the same time ARPA continued to invest heavily in rocket development, albeit with military rather than scientific intentions. In 1957, it actually considered establishing a dedicated university-affiliated aerospace laboratory, akin to the national laboratories but with a space focus.[27] Although that effort came to naught, it planned a substantial program of missile development even as NASA was being created. In 1958, ARPA planners laid out an ambitious schedule for the succeeding decade, including three satellite

launches by 1959 and ten by 1960, lunar probes, orbiting manned capsules, million-pound-thrusting rocket engines, fluorine fuel development,
satellite-born reconnaissance systems, and satellite-launched missiles. The
estimated price tag for the whole development catalogue was $450 million –
roughly five times the proposed NSF budgets for the same time period.[28]

Sputnik had exposed the early gap between Soviet and American rocket
technology. The gap had its roots in decisions made as far back as the late
1940s to focus on perfecting small-to-medium range missiles for military use
rather than on developing large rockets capable of launching payloads into
orbit. DuBridge, writing after the fact, rued that early decision as lacking
"adequate foresight" but admitted that it seemed reasonable at the time.[29]
And even Sputnik's launching appeared to create a political crisis more than
a tactical military one. By early 1958, however, President Eisenhower had
pivoted. He had aggregated military space systems in ARPA, civilian and
scientific space systems in NASA, and was entertaining ambitious proposals (with equally ambitious budgets) for the two programs over the coming
decade. The United States had started slowly in space but had now committed itself to catching up.

Satellites

Sputnik had sown fear in Americans, but Sputnik II, successfully launched
the following month, convinced the nation that the Soviets were substantially ahead in the space race. The new satellite weighed over a thousand
pounds which meant that the Soviets possessed a rocket capable of generating at least 500,000 pounds of thrust – a threshold which the American
rocket program was unlikely to reach for at least two years.[30] But even more
sobering was that the satellite carried a dog, Laika, the first living being to
enter orbit. Although Laika died within several hours of launch from over-
heating, her mere presence suggested to an increasingly anxious American
public that the Soviets were preparing for manned spaceflight. *Life* magazine, which devoted nearly a third of an issue to the launch, warned, "The
Soviets had burst upon the world as the infinitely sinister front runners in
the sophisticated and perilous science of space."[31]

In the wake of Sputnik II, White House discussions grew concomitantly tense. While Eisenhower had downplayed the importance of the first
Sputnik, the second satellite, coming so soon and so successfully, created a
political crisis. White House planners realized that the nation must rapidly
counter the Soviet success to mollify an anxious public and neutralize hawkish attacks coming from Congressional opponents. An internal directive from
the Office of the Science Advisor warned that it was "vitally important that
the United States launch a satellite at the earliest possible moment and then
go on to achieve spectacular successes in the field to offset the psychological, political, and military impact of the Soviet's rapid advance."[32] Directives
went out to the Navy, well along in its Project Vanguard satellite program,

to "achieve substantial success in the very near future," and that for reasons of "national prestige" every effort must be made to minimize the chance of failure.[33] Beyond the political dimension, there was a growing tactical and strategic need for a functioning satellite program. The Soviets would soon have functional intercontinental ballistic missiles and the United States would need satellites to track and counter the new military technology, as well as for general reconnaissance, intelligence, weather mapping, communications, and geodesy.[34] While all of these needs had been foreseen, the surprising Soviet success made an American response substantially more urgent.

The nation's interest in satellites dated to discussions directly after World War II which had led to competing proposals by the Office of Naval Research and the Air Force's Project RAND in 1947 and 1948 to create a program leading to a satellite launch. RAND suggested that a workable satellite could be built and launched in five years for a total of $150 million, which might "inflame the imagination of mankind and would probably produce repercussions . . . comparable to the explosion of the atomic bomb."[35] The satellite might be used to observe weather conditions, relay radio signals, observe enemy aircraft, and guide missiles and bombing raids. While the DoD's Committee on Guided Missiles rejected both the Navy and Air Force proposals as unnecessary, Secretary of Defense James Forrestal earmarked funds to study the idea.[36]

In 1954, a group led by von Braun of the Army Ballistic Missile Agency recommended moving forward with a satellite development program based on the Army's Redstone rocket.[37] At nearly the same time, NRL was developing its own Viking rocket, capable of carrying a second stage which could possibly carry craft into the upper atmosphere, while the Air Force was working on its own Atlas rocket which could also conceivably serve as a satellite booster. Notably, both the Navy and Air Force proposals were presented as purely scientific in scope while the Army's program, resting as it did on a wartime weapon, appeared to be more closely tied to the military.[38] The President was spurred to act on one of the proposals once the plan for the International Geophysical Year was announced in 1954. In July 1955, the DoD agreed to participate in a project with the intention of producing "scientific observations" of the upper atmosphere,[39] and that September, President Eisenhower ordered NRL to move forward with its Viking-based satellite program, now named Project Vanguard.[40]

A satellite was potentially useful for its ability to hover in the upper reaches of the atmosphere (above 100,000 feet) for a long time. While rockets could go up that high to take pictures and make radio observations, they stayed at that height only for a few minutes before descending and crashing to Earth. In 1949, the Bumper-WAC, a two-stage vehicle built on a V-2 rocket, had gone to 250 miles. Most other efforts had used single-stage rockets and had achieved heights of about 100 miles or less. While such flights have produced important data – air pressure and temperature measurements, along with observations of the ionosphere and solar and cosmic radiation – no flight had lasted more than 16 minutes from launch to reentry, and the time spent in the targeted upper atmosphere was generally less than ten minutes. An orbiting satellite, by contrast, could stay up for weeks or even years. When the United States eventually successfully placed its first Vanguard satellite into orbit, in 1958, it was designed to remain in orbit for 200 years.[41]

Getting the satellite up would require a significantly more powerful rocket than the United States possessed at the time. The nation's most mature rocket technology continued to rest on the old V-2 architecture brought from Germany by von Braun and his colleagues, and while the model was reliable it had been designed to move a warhead from Germany to the United Kingdom: a distance of about 600 miles at a velocity of 2,500 miles per hour. A rocket placing a satellite in orbit would need to approach 17,000 miles per hour at its apogee while fighting the full force of the Earth's gravity the entire way. The effort would require sextupling the power of the conventional V-2, almost certainly necessitating a second stage and possibly a third. While von Braun had made preliminary plans for a two-stage rocket at Peenemünde in 1945 (with the hope of targeting the United States), he had never successfully built one.[42] A *manned* satellite would require yet a larger rocket and, according to a Temple University analysis, "at least a 20-fold [increase in] expense of human effort, money, and time . . . with an inestimable amount of human ingenuity and invention."[43] Such an achievement would constitute a true milestone in human achievement, akin to circumnavigating the globe and conquering the poles. On the other hand, a Soviet victory would be devastating – a "serious blow to the technical and engineering prestige of America the world over."[44]

Satellites could produce multiple benefits but underlying all was reconnaissance. In 1955, President Eisenhower proposed to Soviet premier Nicolai Bulganin that the two countries engage in a policy of "open skies" wherein each country would allow airplanes to fly over their airspace for both commercial and intelligence reasons. When the Soviet premier rejected the notion, the United States began to envision a fleet of reconnaissance aircraft capable of flying at high-enough altitudes to evade antiaircraft fire.[45] A satellite could be more effective, given its regular orbital tracks over the Soviet Union. In 1957, the Army proposed putting into orbit a 500-pound satellite at an altitude of 300 miles with a camera capable of discerning missile launching sights, ships, airports, and industrial installations. Such a satellite would make overlapping sweeps of Soviet territory and would produce images of virtually all Soviet territory once every three days. The satellite would exploit the most contemporary television recording technology coupled with magnetic recording tape, and would take a series of photos of swaths of territory ten miles wide by 100 miles long. Powered by solar cells, the satellite would have the capability of "full photo reconnaissance" of the Soviet Union.[46] The launching of Sputnik, of course, only heightened the urgency of the program. In 1959, White House intelligence staff starkly argued the case for the satellite program: "Reconnaissance information from over flights of the U.S.S.R. is now vital for U. S. security – to insure adequate order of battle, economic data, warnings of surprise, and the like."[47]

Satellites could also aid in radio communication. Shorter radio waves were impractical over longer distances since they could not bend over the horizon. A geostationary satellite could either bounce signals to a receiving station (passive relay) or actually receive the signal, amplify it, and retransmit it (active relay). A passive relay was technically easy to make and drew no power in its communication function but required a fairly large antenna to receive and bounce the signal; one model proposed in the late 1950s used ground receiving dishes that were 250 feet in diameter and a satellite-born dish 100 feet across. The dish in this case would be an extremely thin sheet of reflective Mylar unfurled across a frame once the satellite was at its orbital altitude of 22,000 miles. Satellites in lower orbit could be used as well, with smaller dishes, but would need to be placed closer together with a series of ground-to-satellite bounces – some 20 satellites to propagate a signal 5,000 miles. One problem with a passive communications satellite is that anybody could use it; it would be essentially a floating wall off of which anybody with a powerful enough transmitter could bounce radio waves.

An active relay satellite would be technically more complicated but could work with weaker ground signals since it would actively amplify the signal upon reception. It could be controlled by the nation of origin and would probably be more reliable. Moreover, an active system could actually store transmission signals on magnetic tape and then relay them later when weather conditions were clearer. Such a satellite, however, would be substantially heavier – a minimum of 1,000 pounds for a 100-watt transmitting satellite. It would draw more power and require some advances in vacuum tube development to increase reliability and decrease weight.[48]

Useful as satellites could be, however, White House planners tended to frame the satellite project as fundamentally an aid to science. A satellite orbiting above the Earth's atmosphere was capable of capturing light and radiation which would otherwise be scattered by the Earth's ionosphere and the heavier gasses below. A satellite-mounted telescope would be able to view stars and planets much more clearly than even the most powerful Earth-bound instruments. And satellite-born radiation sensors would be able to detect and map the huge amount of ultraviolet and ionic radiation emanating from the sun which often disrupted long-range radio communications on Earth. Instruments on a satellite would be able to detect cosmic rays, geodesic aberrations, and geomagnetic fields. In the future, infrared lenses might be able to map temperature patterns on the Earth's surface and calculate more accurately the portion of solar energy ultimately trapped as heat. In the distant future, a satellite orbiting hundreds of thousands of miles from the Earth might be able to observe changes in the resonant frequency of an atom proving the existence of Einstein's theorized "red shift."[49] Biological data could be produced as well, with the launch of mold and yeast cultures, and eventually small mammals.[50]

President Eisenhower announced Project Vanguard to the public on July 29, 1955, and set a target date for the first satellite launch of December 31, 1958. The booster rocket would need to lift the satellite to an altitude of 300 miles and accelerate it to nearly 17,000 miles per hour. The task would require a three-stage rocket and would be overseen by NRL, with assistance from the Army and the Air Force. The Navy contracted with the Glenn L. Martin Company of Baltimore to construct the Viking rocket which would constitute the first stage of the booster. In charging the Navy with the task, President Eisenhower opted for a fully American rocket program, which appeared to be far along in development. PSAC and the Office of Defense Mobilization emphasized the need for success over multiple launches to secure American prestige.[51] The initial budget for the entire project was $110 million.[52]

The Vanguard rocket assembly, as designed, stood seven stories tall and weighed 11 tons. Its first two stages burned for just over two minutes, elevating the third stage and satellite into a long, graceful arc. At an altitude of 300 miles the third stage, a solid fuel booster, ignited and burned for only 30 seconds – enough to push the satellite contained in its nose to near escape velocity, at which point a powerful spring pushed the satellite away from the third stage effectively creating two orbiting satellites.[53] While the Martin and Grand Central Rocket Companies built the boosters, NRL created the onboard instrumentation, designed the actual satellites, and coordinated the launch and tracking of the missions.[54]

Although a successful launch became a top priority after the Sputnik launches, the project never quite commanded the attention of senior Pentagon officials. Although controlled by the Navy, it was essentially a non-military venture (albeit with long-term potential military applications). Moreover, because Eisenhower had bypassed the Army's Redstone-based Project Orbiter proposal, the Vanguard project became a somewhat divisive point for the Pentagon. In May 1956, the project was denied "S" status (top priority), and the DoD actually paid for the program through its emergency fund rather than through a dedicated budget line. And while the national committee for the International Geophysical Year recommended constructing six satellites, the Pentagon balked at the additional cost and refused to expand the program. Ultimately, the Army prevailed in having its Project Orbiter named as a back-up program, based on a retooled V-2 rocket now renamed the Jupiter-C, but the decision had the unintended effect of actually diverting attention and funds from Vanguard.[55]

Blasting a satellite into the sky atop a three-stage rocket was risky. The two liquid stages contained tons of liquefied oxygen, hydrogen, and nitric acid, all stored at extremely cold temperatures and fantastically flammable. Each stage needed to ignite and burn at precise intervals, and leaky gaskets or cracked tubes or metalwork could produce a catastrophic explosion. During the third-stage burn, the satellite needed to be guided precisely to a correct altitude and speed to establish a sustainable orbit. Advisors on

PSAC estimated in 1957 that any given satellite launch had only a 50 percent chance of success and recommended that 12 missions be planned to successfully place six satellites into orbit.[56]

Intense pressure from the administration to counter the two Sputnik launches led the Navy to launch a Vanguard rocket on December 6, 1957, that burned for only a few seconds before toppling over. The failure was embarrassingly public – the most "miserably spectacular fashion possible in terms of the level of national aspiration which the press had built up," in the words of one internal Navy assessment.[57] Over the next year, two more Vanguard missions (out of total of six attempted) also failed to place a satellite into orbit. Drew Pearson, a widely read columnist, wrote that the public was getting

> so accustomed to hearing that Project Vanguard has failed to get off the ground that they scarcely pay any attention to the routine announcement that "the second stage failed to ignite," with another 20-inch satellite going to the bottom of the Atlantic.[58]

Historian Enid Curtis Bok Schoettle recalls the nicknames given by the public to the failed Vanguard efforts: "Puffnik, Flopnik, Phutnik, Kaputnik, and Stayputnik."[59]

Shortly after the failure of the first Vanguard launch (paradoxically never referred to as Vanguard I, but rather as TV-3), von Braun and the Huntsville team working on the Orbiter project convinced Eisenhower to let the Army try. On January 31, 1958, the Army successfully placed an Explorer satellite into orbit. Explorer I was small – only about the size of a grapefruit – and the public quickly compared it negatively to the half-ton Sputnik II, which had been large enough to carry a living dog. The Orbiter team, led by the Army Ballistic Missile Agency, coordinated the launch with a California Institute of Technology team based at the Jet Propulsion Laboratory in Pasadena. Like the Navy, the Army had used a number of private sector contractors to build the launch vehicles, among them Chrysler, Ford, North American Aviation, and Reynolds Metals.[60] Over the following year, the Army attempted to place four more Explorer satellites into orbit, succeeding with two.

The Navy continued to attempt Vanguard launches, ultimately succeeding on March 17, 1958, with Vanguard I, and then with three more over the following year. By October 1958, three American satellites were still in orbit while only the last (Sputnik III) remained so.[61] The program had met the threshold of success originally defined and predicted by NRL's development team two years before. Upon questioning from the House Appropriations Committee, the Secretary of Defense defended the program as "a highly successful one in comparison with other large rocket development programs."[62]

Despite the shaky start to the American satellite program, it quickly exceeded its goals. From the earliest small Explorer units, American satellites were able to reveal previously invisible currents in the Earth's magnetic field and more accurately map the planet's asymmetry.[63] Vanguard units produced data concerning efficacy of geocentric orbits, life expectancy of cells in zero gravity, and the impact of micrometeorites.[64] By 1963, American satellites were routinely providing navigational data to the Army and Navy, mapping Soviet military installations, and aiding with meteorological forecasts. The United States was launching over 50 satellites annually and by the end of that year had 248 in orbit. Plans to simulcast the 1964 Olympics using satellite television relays were in place and new rocket engines were under development informed by the successes and failures of the Redstone, Jupiter, Atlas, Viking, and Vanguard boosters.[65] And most satisfying, American satellites stayed up, as opposed to Soviet satellites which quickly fell down. John Hagen, the director of Project Vanguard, proudly boasted in 1958:

> We know now that if Benjamin Franklin, whose science was most modern for his day, put Vanguard in orbit it would still be circling the Earth and we today might be speculating on when it might return to the atmosphere. It will take long observation to make possible a more accurate prediction, but we know now that Vanguard will be with us for at least 200 years.[66]

Research in the Space Age

The military branches continued to dominate the nation's research through the late 1950s. Just about half of the nation's entire research budget of $5.4 billion came from the federal government, and about 90 percent of the federal investment went to defense projects (including work done for the AEC). Work in aviation and electronics dominated, although defense-related research also included work in communications, instrumentation, metallurgy, and chemical engineering.[67] Of this, no more than 8 percent went to basic research (although this was difficult to clearly define), and even here more came from the DoD and the AEC than the NSF – the agency actually charged with funding the nation's basic research. Universities conducted $460 million of research of all types (most of this funded from government sources) of which only $160 million went toward basic research. The national laboratories, along with the Jet Propulsion Lab, the Advanced Physics Lab, and Lincoln Lab drew about half of all of this funding, somewhat obscuring the size of the university engagement with defense work.[68]

The continued scope of the military research budget, coupled with coordination problems surrounding the Vanguard mission, led the White House in 1957 to create a new Assistant Secretary of Defense for Research Engineering to take over the work of the old Research and Development Board (RDB) and coordinate the work of each branch's research bureaucracy.[69] The recommendation for the reorganization had come out of a commission chaired by former President Herbert Hoover, which recommended that the many components of DoD research – nuclear weapons, missiles, space technology – be brought together under one administrative portfolio, even as each agency maintained some independence in developing its own weapons systems. The new assistant secretary would need to think strategically about new types of combat such as undersea warfare, aerial continental defense and, most urgently, space-based weaponry and reconnaissance. More importantly, the new office would serve as a critical liaison in binding defense needs to the civilian research sector. A PSAC memo to President Eisenhower in 1958 stated,

> It is of great importance that we have a research and development organization that can strike deep roots into our civilian scientific community and can tap our most basic and advanced research so that military research and development can rapidly take advantage of new discoveries in basic research which may be profoundly influential in new weapons technology.[70]

Increasing demands for highly sophisticated new weaponry, born of advanced avionics and satellite transmission, led Secretary of Defense Neil McElroy to create the Defense Special Projects Agency in 1958, later renamed the Advanced Research Projects Agency (ARPA). Herbert York, the director of the Livermore National Laboratory, became ARPA's first director, which was largely an administrative response to the embarrassment of Sputnik and the two Vanguard failures. Although the enabling executive order was deliberately vague (the operational clause simply read that "the Agency is authorized to direct such research and development projects being performed within the DoD as the Secretary of Defense may designate"), the new agency would actually be quite powerful. It would usurp the individual services' work on missiles, satellites, and space flight and bleed resources from their budgets. Historian Sharon Weinberger writes that the Pentagon's reception to ARPA was "ice cold . . . a threat to their turf, and their budgets."[71]

ARPA was designed to work closely with universities, or at least with university stepchildren such as JPL, APL, Lincoln Labs, and the new Institute for Defense Analysis (IDA) at MIT. Working through the Weapons Systems Evaluation Group, the DoD contracted with the IDA to test and evaluate prototypes and intermediate versions of evolving weapons. IDA, in turn, invited Caltech, the Case Institute of Technology, Stanford, and Tulane to help create a consortium to staff the IDA and contract with the DoD.[72]

Continuing his government realignment, President Eisenhower created NASA later that year, under which he placed the functions of the old NACA and simultaneously moved nearly all of the Defense Department's missile and rocket development work, including Project Vanguard and projects which included satellites and lunar probes. As had been true with ARPA, NASA expropriated substantial pieces of the DoD research and development budget, including (explicitly) $59 million of ARPA work and nearly the same amount of Air Force work.[73] The new agency quickly grew to an enormous size, employing within five years more scientists and engineers than any other government agency. In addition to working closely with JPL, contracting with many independent university-based scientists, and employing thousands of engineers directly in its own research operation, it owned and operated the launch facilities at Cape Canaveral, the Houston control center, and over a dozen radio tracking facilities around the world.

NASA was deliberately cordoned off from the military at its creation in hope of framing America's space mission as scientific and civilian. At the same time, it was created in an environment of ferocious military competition with the Soviet Union, in which space flight served as a proxy indicator of a nation's military missile and navigational capability. The new agency was thus not only bonded to the military, contravening Eisenhower's vision, but in fact it was spurred from its onset to quickly counter Soviet space progress. Many scientists and planners rued the fact that the space race impelled NASA to develop and launch rockets and missions which were overly derivative of existing military equipment and insignificantly sophisticated to optimize mission-specific technology. That is, NASA scientists lacked the time to ideate in innovative ways. Its work was pressured to the "spectacular" (in the words of Lee DuBridge), and it was forced to bypass work which might have been more scientifically constructive if less appealing to the public.[74] And while the spectacular flights of the Mercury, Gemini, and Apollo programs captured the public's imagination and won political support, they did so at the expense of more thoughtful science. DuBridge would write that in matters of space exploration, purely scientific considerations would "not be the only, or even the predominating, factors in the formulation of government policy."[75]

Fascination with space delayed a rebalancing of basic and applied research and of military and civilian research. The hundreds of billions of dollars thrown into the manned space program over the following decade was nearly all in applied engineering fields, and thus even as the NSF budget grew during that time it continued to lag behind applied federal science funding. The NSF budget had, in fact, continued to grow since its inception in 1950. It was given $40 million in 1956 and $65 million in 1957 (in contrast to its first Congressional allocation of $250,000).[76] Although this reflected budgetary growth of 50-fold over less than a decade, the initial allocations had been risibly inadequate, and the foundation's budget would need to continue to grow quickly were it to play the dominant role it had been designed to play in the nation's basic research enterprise. In 1963, President Kennedy requested an increase in the NSF budget to $589 million, which was cut by the appropriations committees by nearly half. It would not reach the level set by Kennedy until 1970.[77]

Part of the barrier in growing the NSF budget was the persistent problem of the concentration of the funds. Over a decade into the work of the foundation, the preponderance of funds continued to flow to a handful of research behemoths located disproportionally in the northeast and in California. In budget hearings, Representative Albert Thomas (D., TX) demanded to know why Lee DuBridge, the president of the California Institute of Technology and a member of the National Science Board, continued to divert funds toward his own institution, despite it being supported by funding many times the size of funding allocations at most of the nation's research universities. "You can tell the doctor (DuBridge) to go to other fields. He has already conquered this one," admonished the Congressman to the President's science advisor, Jerome Wiesner. "I do not think any particular state has a monopoly on intelligence."[78]

Through the 1960s, despite recommendations that basic research be raised to 15 percent of the government science funding effort, the true budget never came close.[79] Rather, the combined research and development budget of the Defense Department and of NASA grew so rapidly during the decade that even the small portion of these funds diverted to basic research overshadowed the total size of the NSF. "It goes up so fast that I cannot follow it," admitted Senator William Fulbright (D. AR) in 1969. In that year, the DoD alone spent $7.2 *billion* on research and development, of which about 8 percent, or $560 million, went to projects which could reasonably be called basic or non-applied – a touch greater than the total NSF budget.[80] In that year, too, the NASA budget drew about $4 billion of federal funds, although here it was harder to break out basic research given the applied nature of the majority of funds. By any measure, the NSF effort paled in comparison. A heartfelt commitment to basic science simply could not raise the enterprise to a status equal to Cold War military concerns or to the space race, which had so captured the nation's attention.

Test Bans and Disarmament

The age of satellites shocked Cold War adversaries into considering nuclear disarmament or at least slowing the rate of accumulation of new weapons. Concern over nuclear Armageddon was widespread; as a US Senator in 1959, John F. Kennedy remarked that the United States and the Soviet Union were in a position "to exterminate all human life seven times over" and that the ultimate price of the arms race was "death – for both."[81] A White House working group chaired by Cornell physicist Hans Bethe, with input from representatives of the DoD, the AEC, the CIA, and PSAC, apprised President Eisenhower in a report of March 1958, that the United States could safely cease nuclear testing without jeopardizing its security.[82] Following the report, President Eisenhower invited Soviet Premier Nikita Khrushchev to join the United States and the United Kingdom in a conference to explore ways in which participating nations could verify the presence or absence of testing activities. In late August, the nations met in Geneva.[83] Concerns over the seismic "footprint" of nuclear tests and the need for on-site inspections to verify the absence of testing largely scuttled the negotiations. The talks were further undermined by a 1959 panel report authored principally by Robert Bacher of the California Institute of Technology which argued that even on-site inspections would likely fail to discern nuclear tests. The Soviet foreign minister, Andrei Gromyko, put a final end to the negotiations two years later.[84]

A separate but related movement advocated banning atmospheric tests of nuclear weapons which were poisoning the environment with radioactive fallout. Nuclear testing emitted radioactive iodine (I^{131}), strontium (Sr^{90}), and cesium (Cs^{137}). Fallout from Soviet tests was spreading across the northern hemisphere and producing high concentrations of I^{131} in milk. While the iodine decayed rapidly, lower amounts of radioactive strontium and cesium remained longer in the atmosphere and then decayed more slowly once falling to Earth. They could be detected at elevated levels in meat and dairy products for months after the test explosions.[85] Through the 1960s, atmospheric nuclear testing (largely Soviet) doubled the levels of Strontium in food supplies and tripled strontium levels in human bone.[86]

Eschewing atmospheric testing, the United States turned to underground testing in 1963, conducting numerous tests at various depths and in differing rock formations over the following decade, particularly in the Nevada desert. Such tests tended to trap radioactive material underground where it was sealed under the tons of falling and cratered rock which the nuclear blast created. The Federal Radiation Council, which monitored and set standards for testing, stated in 1962 that underground testing produced risks that were "so slight that countermeasures may have a net adverse rather than favorable effect on the public's well-being."[87] Over the next six years, the nation conducted tests through its Boxcar, Benham, and Sulky, and Palanquin testing series, all burying their own nuclear waste under hundreds of feet of debris.[88]

The potential for satellite-based weaponry, or the reality of missile-delivered bombs, led the two superpowers to consider proposals for dedicating space exploration to peaceful purposes. The problem was that long-range missiles were already capable of delivering nuclear warheads with great accuracy over long distances at speeds of 3,000 miles per hour, and unilateral shedding of this technology would lead to a strategic imbalance. While both sides could, and did, agree to limitations on satellite-born weaponry, this agreement was undermined by the very inefficacy of satellite-based technology. Weapons-bearing satellites required enormously powerful rockets to place in orbit and were vulnerable to attack from an opponent. Far more dangerous were the dispersed siloed missiles, which the two sides aimed at each other, and the growing fleet of missile-carrying nuclear submarines, which were undetectable and could stay underwater indefinitely. Jerome Wiesner concluded that satellites represented a "poor approach to long lived, confident weapon systems for a deterrent strategy."[89]

Kennedy persevered in efforts to limit nuclear proliferation. Weapons grew larger and more destructive by orders of magnitude; ICBMs could now hit targets a continent away in arcs of trajectory making them nearly impossible to intercept and defend against. Secretary of State Dean Rusk explained the desperate need for a comprehensive and enforceable agreement between the leading nuclear powers:

> A single thermonuclear weapon can carry the explosive power of all the weapons of the last war. In the last war they were delivered at 300 miles per hour; today they travel at almost 300 miles per minute. Economic costs skyrocket through sophistication of design and by accelerating rates of obsolescence.
>
> Our objective, therefore, is clear enough. We must eliminate the instruments of destruction. We must prevent the outbreak of war by accident or by design. We must create conditions for a secure and peaceful world.[90]

In 1962, Kennedy pledged to end atmospheric tests provided that the Soviet Union ceased as well.[91] The pledge, however, was not implemented, and over the following three years American and Soviet atmospheric testing continued. The Soviets justified their stockpiling and testing as a response to the "imperialist aggression." In 1966, *Pravda*, the official news organ of the Soviet State, dismissed American disarmament promises as rhetorical hypocrisy and committed the Soviet Union to nuclear parity. "The USSR is ready to live in peace with all countries, but it will not put up with the imperialist arbitrariness," an editorial read.[92] A 1967 effort to negotiate a broad nonproliferation treaty failed. The two countries would not sign a treaty until 1968, and that treaty would not take effect until 1970.

Test bans and arms control treaties affected American research policy only peripherally through the 1960s. DoD, AEC, and now NASA funding grew substantially over the decade, with work turning to particle physics, missile technology, avionics, rocketry, and space exploration. Sound research rarely relied on weapons production or atmospheric testing. Interest in weapons performance was focused on underground testing and the potential to use nuclear arms toward peaceful ends.

The more pressing question of the period was the ultimate shape of space research. Would space simply be an extension of sea and air – another venue in which to carry on war and war-related research and development? Or would the new frontier override security concerns in the interest of exploration and basic science? Control of space had military consequences, and NASA was never quite able to shed its origins in Army and Navy aviation. Early astronauts would be drawn from the pool of gifted test pilots who populated Air Force and Navy training bases, and NASA's technical roots lay in NRL. But space did not lend itself easily to militarization, and excitement over the frontier seemed to transcend Cold War concerns. While "conquering" space would provide bragging rights well into the 1970s, those boasts would remain just that – rhetorical flourishes with little practical application in the Cold War of science and armaments. Space science would always be associated with the military but never wholly militarized. Eisenhower's repeated defense of the American space program as being fundamentally scientific and peaceful was neither wholly honest nor entirely cynical. Achievements in space projected technical mastery and suggested military prowess. But the suggestion was just that – a suggestion.

Notes

1 Enid Curtis Bok Schoettle, "The Establishment of NASA," in Sanford Lakoff, ed., *Knowledge and Power: Essays on Science and Government*, Free Press, 1966, pp. 173–74.
2 The Smythe quote is from Lloyd Berkner, "Wanted: A National Science Policy," *Atlantic*, 201, January 1958, p. 40. The Adams quote is from Charles Maier's introduction to George Kistiakowsky, *A Scientist at the White House*, Harvard University Press, 1976, p. xxx.
3 Grazhdanskaia Aviatsiia, "New Era in the History of World Science: Soviet Moon," November 10, 1957, RG 255, NASA/Vanguard, 3: Reference File, p. 6.
4 Eisenhower, "Address to the Nation," November 7, 1957, AW, 26: NSF 1957, p. 1.
5 The Polaris missile program, built upon the Army's existing Jupiter program, began operations with the SSBN nuclear submarine *George Washington* in 1960. Submarine-based nuclear-tipped missiles became a central component of the nation's Cold War deterrence strategy. Polaris missiles were eventually replaced by the multi-warhead Poseidon missiles, which were augmented in the 1980s by the Trident program.
6 The United States never developed a supersonic bomber program.
7 Eisenhower, "Address to the Nation," November 7, 1957, p. 3.
8 Lloyd Berkner, "Wanted: A National Science Policy," p. 40.
9 Quoted in Schoettle, "The Establishment of NASA," p. 174.
10 Eisenhower, "Address to the Nation," November 7, 1957, p. 6.
11 Quoted in Maier, *A Scientist at the White House*, p. xxix.
12 Eisenhower, "Address to the Nation," November 7, 1957, p. 8.
13 Notably, more than 9,000 slave laborers drawn from Nazi concentration camps died in producing the V-2s. A few of the rockets were also launched toward Lille, Paris, and Liège.
14 The reference is from Walter Dornberger, *V-2*, Viking Press, 1954.
15 American Rocket society, "Space Flight Program: Report by the Space Flight Technical Committee," October 10, 1957, OSAST, 15: Space (2), pp. 4–5.
16 Ibid., p. 8.
17 A satellite placed into orbit at that altitude circles the Earth once daily. Because the Earth also rotates on its axis once daily, the effect is that the satellite remains motionless above the Earth.
18 S.P. Johnson and E.M. Purcell, "The Scientific Exploration of Outer Space," March 7, 1958, OSAST, 15: Space (4), p. 7.
19 Ibid., p. 10.
20 Ibid., p. 18.
21 Quoted in Von Hardesty and Gene Eisman, *Epic Rivalry: The Inside Story of the Soviet and American Space Race*, National Geographic, 2008, p. 63.
22 White House Press Release, February 5, 1968, OSAST, 15: Space (4).
23 Maier, *A Scientist at the White House*, pp. liii–lix.
24 White House Press Release, October 1, 1958, RG 255, NASA/Vanguard, 1: Background.
25 Audra Wolfe, *Competing with the Soviets*, Johns Hopkins University Press, 2013, p. 92.
26 Lee DuBridge, "Policy and the Scientists," *Foreign Affairs*, April 1963, p. 582.

27 "Space Weapon Concept Discussed at Secretary of Defense Luncheon," November 1, 1957, OSAST, 15: Space (2).

28 Herbert York, "ARPA Space Program," March 17, 1958, OSAST, 15: Space (4).

29 Lee Dubridge, "Policy and the Scientists," *Foreign Affairs*, April 1963, p. 583. Hans Bethe disagreed with the assessment, suggesting that US and Soviet rocket technology were on equal footing, albeit with different emphases: the United States on ICBMs, and the Soviets on manned space flight. See, Bethe to Citron, June 26, 1962, JFK, WHCF, OS:'61-'62.

30 The estimate was from Senator Stuart Symington (D., MO), a member of the Armed Forces and Foreign Relations Committees, but was echoed widely by the media. Symington's estimate carried particular weight, as he had served previously as the first Secretary of the Air Force (1947–1952). See Robert Divine, *The Sputnik Challenge*, Oxford University Press, 1993, p. 43.

31 Quoted in Ibid., p. 44.

32 "Briefing on Army Satellite Program," November 19, 1957, OSAST, 15: Space, (2), p. 1.

33 "Justification of Estimates," RG 255, NASA/Vanguard, 1: Background, p. 2.

34 "Briefing on Army Satellite Program," November 19, 1957, OSAST, 15: Space (2), p. 1.

35 Quoted in Von Hardesty and Gene Eisman, *Epic Rivalry*, p. 57.

36 Quoted in Bok Schoettle, "The Establishment of NASA," p. 166.

37 Von Braun and nearly 1,600 of his associates who worked on the German V-2 rocket program were brought to the United States immediately after World War II, through the Army's Project Paperclip. For details, see Linda Hunt, *Secret Agenda: The United States Government, Nazi Scientists, and Project Paperclip, 1945–1990*, St. Martin's Press, 1991.

38 The interservice competition for the satellite project, which pitted the Army's Redstone rocket against the Navy's Viking missile is clearly documented. The exact reason for the President's favoring the Navy project is unclear. I follow here Bok Schoettle's interpretation of the events. See Ibid., pp. 166–170.

39 Satellite Project Fact Sheet, July 29, 1955, RG 255, NASA/Vanguard, 2: Press Release, p. 2.

40 Although the satellite project was framed as scientific in nature, even the earliest documents include military applications of the project. A 1955 internal National Academy of Sciences memo lists "observations of Soviet satellites" as the first possible use for an American satellite. National Academy of Sciences/National Research Council, Odishaw to Porter, November 16, 1956, AW, 27: NSF, 1957.

41 Satellite Project Fact Sheet, July 29, 1955, p. 1.

42 Research Institute of Temple University, "Report on the Present Status of the Satellite Problem," August 25, 1953, DDE, NASA, 1: Status of Satellite Program, 1953, p. 3.

43 Ibid., p. 4.

44 Ibid., p. 6.

45 Open Skies was eventually successfully negotiated by President George H. W. Bush in 1989, and ratified in 1992. In 2002, the Treaty on Open Skies was ratified by 34 nations. The United States withdrew from the consortium in 2020. Notably, President Bush had previously been the director of the CIA and entered office with substantial awareness of the nation's intelligence needs.

46 "Briefing on Army Satellite Program," November 19, 1957, OSAST, 15: Space (2), p. 6.

47 R.S.L., "Political Action and Satellite Reconnaissance," April 24, 1959, OSAST, 15: Space (6), p. 1.

48 PSAC, "Communications Using Earth Satellites," no date, OSAST, 13: Space (8).

49 J.W. Joyce, "The Scientific Aspects of the Artificial Earth Satellite," AW, 27: NSF, 1957. Also, John Hagen, "Scientific Research with Artificial Satellites," April 28, 1958, RG 255, NASA/ Vanguard, 4: ARPA.

50 American scientists generally viewed the launch of Laika as more a publicity stunt than a serious biological research enterprise. Hiden Cox, the director of the American Institute of Biological Sciences felt that launching yeast cells would yield "infinitely more significant data" than putting up a dog. From a wire release, January 22, 1958, RG 255, NASA/Vanguard, 2: Press Release.

51 Rabi to Flemming October 10, 1956, RG 255, NASA/Vanguard, 1: Background.

52 Livermore to Klaisdell, June 21, 1957, RG 255, NASA/Vanguard, 1: Background.

53 Data from a Martin press release, "Navy-Martin Vanguard Satellite Launching Rocket," RG 255, NASA/Vanguard, 4: ARPA.

54 NRL, "Project Vanguard," RG 255, NASA/Vanguard, 1: Investigation.

55 Bok Schoettle, "The Establishment of NASA," p. 170.

56 York to Killian, December 19, 1957, OSAST, 12: Interdepartmental Committee on Satellite Research.

57 "Real problem is . . . "June 29, 1958, RG 255, NASA/Vanguard, 4: ARPA, p. 1.

58 Drew Pearson, "U. S. Still Losing in Space Race," *Washington Post*, 1958.

59 Bok Schoettle, "The Establishment of NASA," p. 194.

60 "Teamwork Produced Explorer IV," RG 255, NASA/Vanguard, 2: Press Release.

61 Notably, the Soviet satellites were substantially larger than the American ones; Sputnik III, for example, was 80 times the weight of Explorer III. See Robert Divine, *The Sputnik Challenge*, p. 185.

62 Answers of Secretary of Defense Neil McElroy to queries by the House Appropriations Committee, Director of Surveys and Investigations, "Statement Probability of Success and Reliability," RG 255, NASA/Vanguard, 1: Investigation, p. 1.

63 John Lear, "The Moon That Refused to be Eclipsed," *Saturday Review*, March 5, 1960, pp. 45–47.

64 "Project Vanguard: Contributions to the Scientific and Space Research Fields," RG 255, NASA/Vanguard, 1: Investigations.

65 NASA, *Annual Report*, 1964, chapter 1.

66 John Hagen, Address for the American Society of Newspaper Editors," April 19, 1958, DoD Press Release, RG, 255, NASA/Vanguard, 2: Press Releases, p. 4.

67 Chester Barnard, "A National Science Policy," *Scientific American*, 197:5, November 1957, pp. 45–46.

68 Ibid., p. 48.

69 Of note, the position was initially an assistant secretary, then the director of Defense Research and Engineering, and then, as of 1977, the Under Secretary of Defense for Research and Engineering.

70 Killian to Eisenhower, January 28, 1958, OSAST, 1: DoD 1957, p. 7.

71 Sharon Weinberger, *The Imagineers of War: The Untold History of DARPA, the Pentagon Agency that Changed the World*, Alfred A. Knopf, 2017, p. 37.

72 Skolnikoff later became a professor of political science at MIT. For more on the IDA, see James Killian, *The Education of a College President*, MIT Press, 1985, pp. 76–78. See also, DoD Press Release, "Dr. Herbert York Named to Head New IDA Division and ARPA Science Post," March 18, 1958, RG 255, NASA/Vanguard, 2: Press Releases, and various documents in OSAST, 11: Institute for Defense Analysis.

73 "Transferring Certain Functions From the Dept. of Defense to NASA," October 1, 1958, RG 255, NASA/Vanguard, 1: Background.

74 Lee DuBridge, "Policy and the Scientists," p. 582.

75 Ibid., p. 583.

76 Sidney Hyman, "The President's New Power," *Saturday Review*, 40, February 2, 1957, pp. 40–44.

77 D. S. Greenberg, "Money for NSF: The Odyssey of a Research Agency's Budget," *Science*, 158, October 20, 1967, pp. 357–361.

78 "News and Comment: NSF Hearings," *Science*, 142:3590, October 18, 1963, p. 142.

79 "Propose Powerful Foe for NSF," *Science News Letter*, May 1, 1965, p. 275.

80 "Defense Research Takes Its Licks," *Science News*, 96, August 23, 1969, p. 146.

81 Quoted in Seaborg, *Kennedy, Khrushchev, and the Test Ban*, p. 32.

82 PSAC, Memo for Killian, May 8, 1959, DDE, PSAC, 5: PSAC (2).

83 Senator Albert Gore (D., TN) was particularly pessimistic. Glenn Seaborg, *Kennedy, Khrushchev, and the Test Ban*, University of California Press, 1981, pp. 14–16.

84 "Statement by the President on Nuclear Test Inspection," March 29, 1962, IIR, 72: ACDA.

85 "Expected Fallout from Atmospheric Weapons Tests," April 20, 1962, LH, 6: Dr. Haworth's AEC Files.

86 Federal Radiation Council, "Estimates and Evaluation of Fallout in the United States from Nuclear Weapons Testing Conducted Through 1962, May 1963, LBJ, NSF-Charles Johnson, 31: Nuclear – Fallout, pp. 1–5.

87 Quoted in Seaborg to Bennett, June 27, 1963, LBJ, NSF-Charles Johnson, 31: Nuclear – Fallout, p. 2.

88 For details, see "Analysis of Project Palanquin," no date, LBJ, NSF, 29: Plowshare Events, vol. 1.

89 Wiesner to Kennedy, August 9, 1962, JFK, NSF, 284: OST, 1961–1962, p. 4.

90 Statement by Dean Rusk at the Second Plenary Meeting of the Conference of the Eighteen Nation Committee on Disarmament, March 15, 1962, IIR, 72: ACDA, p. 3.

91 Max Frankel, "New Atom Tests by Soviet Hinted in AEC Findings," *NYT*, July 1, 1962, p. A1.

92 Quoted and translated in an internal memo, *Pravda* editorial, September 1966, IIR, 72: ACDA, p. 3.

10 New Frontiers

Training, Research, and Manpower, 1960–1968

Reorganization in the Kennedy White House

In 1963, Jerome Wiesner, President Kennedy's senior science advisor, noted the extraordinary progress the physical sciences had made over the previous 13 years. The United States was going through a "great science-generated technical explosion," comparable to the Industrial Revolution of the previous century.[1] Progress in nuclear physics, rocket propulsion, radio and radar, and digital computing had been prodigious. The atomic bomb had multiplied the destructive power of conventional munitions by a factor of a thousand, and the hydrogen bomb had multiplied it by another thousand. Ballistic missiles reduced delivery time of explosives from hours (on propeller-driven aircraft) to minutes. Computers were radically increasing the speed of data analysis, and newly developed transistors promised to increase computing capacity exponentially.[2] A senior scientist had been installed at the top echelon of the DoD to guide strategic planning, and the White House had brought scientific analysis and planning into its senor policy development ranks.

The breakthroughs were built on federal science funding, which had grown from $1 billion annually in the late 1940s to $11 billion in 1962. For 1963, President Kennedy requested $14.7 billion in research funding – an increase of nearly a third in just one year. While this massive increase was due, in part, to exponential growth in the space program (whose budget had grown from $500 million in 1960 to $3.6 billion in 1963), the majority had gone to military-focused research.[3]

The expansion in the federal science effort, coupled with the shock of Sputnik in 1957, forced a reconsideration of the government's science advising apparatus. Congress considered establishing a joint committee on science, a scientific accounting office, and a Congressional Office of Science and Technology (COST).[4] The first was seen as a possible solution to general scientific ignorance among the members of Congress, while the latter two proposals would leverage professional staff expertise to audit scientific funding and provide scientifically knowledgeable input into the general legislative process.[5] Such proposals, however, challenged staff expertise on the

DOI: 10.4324/9781003363897-10

various committees and thus were viewed with suspicion. The House Public Works Committee, for example, was staffed with several civil engineers, while the House Committee on Science and Astronautics had an aeronautical engineer and an electrical engineer on staff, along with two scientific specialists detailed from the Library of Congress. Likewise, the Joint Committee on Atomic Energy had its own staff expertise to call on. Members of Congress tended to develop expertise as they served on committees; Senator Clinton Anderson (D., NM) likened the process of learning through legislating to "osmosis."[6] The legislation establishing COST ultimately failed in committee. Most of Congress felt that scientific expertise belonged in the executive agencies or in the White House itself. Science generally attracted congressional interest only insofar as it bore on military matters, and the Library of Congress's research staff was able to answer questions surrounding scientific matters.

Meanwhile, President Kennedy asked Wiesner to reconsider the structure of science advising within the White House. The current arrangement was centered on PSAC, whose chair was the Special Assistant for Science and Technology. The structure had been in place since Eisenhower had appointed James Killian his special assistant in 1957. Killian was followed by George Kistiakowsky in 1959, who yielded to Wiesner upon Kennedy's inauguration in 1961. From time to time, various presidential commissions and planning groups had proposed creating a federal Department of Science under which to place all federal research funding and activities, or a White House Office of Science, akin to the wartime OSRD, to coordinate scientific programs throughout the federal government.[7] Special panels produced a variety of targeted reports throughout the 1950s, most notably the Technical Capabilities Panel Report of 1954, which drew attention to a growing ballistic missile gap between the United States and the Soviet Union, and the Gaither Panel Report of 1958, which questioned the utility of nuclear fallout defense.

In the 1957 arrangement, the special assistant held quasi-cabinet status. He (initially all were men) served full-time within the Executive Office of the President, attended National Security Council meetings, and met regularly with the President and his chief of staff. The advisor's focus tended to change according to military and strategic considerations; during the late 1950s, for example, much of Kistiakowsky's time was absorbed by matters related to the space program. By 1959, however, the need for a coordinating function (in addition to PSAC's advisory function) impelled the White House to create the Federal Council for Science and Technology composed of principals from agencies which worked with, housed, or funded scientific research.[8] The Federal Council did not so much make science policy or keep the President apprised of scientific advances as keep the White House staff aware of the scientific activities going on within the different agencies. That is, it served essentially to thwart redundancy between the various branches of the armed forces, the AEC, the NSF, and the national laboratories.

In 1962, yet another body entered executive service: the Office of Science and Technology (OST), which served to professionalize the disparate advising staff within the White House, and whose full-time director could testify in an official capacity before Congress. The intent of the dual bodies (OST and PSAC) was to create an external, formal science policy group to report to Congress and the public and a confidential (and unofficial) inner group which could speak in confidence to the President and his senior staff. Wiesner, somewhat bewildered at the range of science policy and advising groups within the executive branch, concluded that the unwieldy apparatus was the natural result of a diffuse and decentralized scientific enterprise driven by individual investigators and their teams. Science seemed to progress best when given broad latitude and allowed to flourish in a variety of agencies and institutions. Tracking and guiding the disparate activities of scientists – many of whom relied on federal funds to do their work – required the work of several supervisory teams and groups answering to the President, Congress, and the President's senior staff.[9]

In a broader sense, the multiple efforts to reorganize science advising and science policy-making reflected an underlying tension over scientific control and centralization. Scientists did their most innovative and creative work when left alone and unhindered, albeit supported by a generous flow of federal funds. But such decentralization invited redundancy (with multiple scientific groups replicating each other's work) and balkanization, in which no single administrator coordinated the disparate scientific working groups working on large scientific problems as had taken place at Los Alamos and the Radiation Lab during World War II. Fights over a unified Assistant Secretary of Defense for Research and Development (as opposed to separate assistant secretaries for the Army, Navy, and Air Force), or a federal Department of Science, were really fights over how tightly the federal funding masters should dictate, or at least coordinate, the work of disparate scientists.

In 1962, Eugene Skolnikoff, a staff member in Kennedy's PSAC and a professor of political science at MIT, suggested that the "vitality" and "general strength" of American science was a legitimate concern of the government and that overseeing the effort required a diffuse set of oversight agencies and analytical functions. Science advisors needed to meet regularly with senior White House staff members, testify before Congress, confer with scientific and university leaders, negotiate with government science administrators, and keep track of scientific developments by attending conferences or reading newly released reports and articles. They needed to balance political and scientific considerations and weigh recommendations against military threats and international tensions. Skolnikoff wrote in an internal 1962 memo:

> To do this requires understanding the technical situation and the political situation, being able to identify the pertinent technical facts or, more often, estimates, and carrying out the analysis in terms of the political

choices the policy-maker must face. This is not the job of the "expert" as usually defined . . . This is the job of someone who is continuously involved in the policy discussions, who has the confidence of the President, who is able to understand the issues and subtleties of both sides of the subject – the technical and the political – and who has the capacity to present the technical situation as an intimate, dynamic part of the issue.[10]

Wiesner was well aware of the challenges of the role, having spent nearly three years advising President Kennedy. Kennedy, an educated and erudite man, had been largely ignorant of science. Nonetheless, Wiesner had found him to be intensely curious in the years in which the two men worked together and invested in mastering the underlying ideas of science even if he could not grasp specific details. Kennedy had sought out Wiesner's counsel while serving as a Senator from Massachusetts and had asked Wiesner, then a professor at MIT, to meet with him semi-regularly to elucidate scientific issues which he confronted. Wiesner wrote of the President:

Many things impressed me then and drew me to him. There was, of course, his charm; but there was much more. I was most impressed by his quick, almost instinctive understanding of problems once he was given the facts. His background ill prepared him for an interest in scientific matters, yet his interest was lively. He was, in fact, then a member of the Harvard University Astronomy Department Visiting Committee. Obviously unprepared to understand the theory of scientific subjects, he tried to get a physical feel of the matter. For example, he was forever trying to get someone to explain electromagnetic propagation comprehensibly. . . . He wanted to know how radio worked . . . why and how did nature really allow energy to be sent through space.[11]

A constant (and discomfiting) issue in the science advising apparatus was its homogeneity. Of the 18 eminent scientists who served on PSAC, the many more who served as staff researchers and advisors to the Special Assistant, and the additional 17 "consultants-at-large" who served as a reserve resource on which PSAC could draw, nearly all came from a handful of the nation's top research universities. As of 1962, nearly 70 percent of the men and women who served in the senior ranks of science advising and policy-making had received their graduate training at just five universities; over the previous decade, and nearly 80 percent had been associated with just eight universities. Few of these advisors (under 20%) were from the life sciences, and only one (Skolnikoff) came from a social science discipline. None came from the behavioral sciences of psychology or sociology.[12]

Such homogeneity was hardly surprising, given how premier scientists tended to congregate at the most eminent departments. But advice drawn from such a patrician pool could hardly help but exclude issues of teacher training, dispersal of scientific knowledge, and general efforts to broaden the public's knowledge of and support for science. Yet these more populist concerns were hardly ephemeral. Science funding depended on congressional support, which in turn responded to voter sympathy and preference. A scientifically ignorant electorate could hardly be expected to support multi-billion-dollar research programs of which it had no understanding. Popular support for science tended toward the showy and fantastic (space travel and satellite launches) or the practical and militaristic (hydrogen bombs, jet planes, ballistic missiles). Mulling over the situation in 1963, Wiesner admitted that an astute science advisor "knows that the moon money could be better spent on scientific projects" yet also knew that Congress would not support them. "The scientists think you are a tool of the administration," Wiesner explained, "and the administration thinks you are a tool of the scientists."[13]

Teaching, Training, and Shortfalls of Talent

Concerns about the volume and quality of the nation's scientific talent had not ebbed. While the supply of scientists and engineers had increased over the previous ten years, so had demand for their services. In 1963, the United States harbored approximately 1.3 million professionally trained scientists and engineers, of whom 500,000 were actively engaged in research and development. Of these, 320,000 were engaged in federally funded projects (mostly based in universities and industrial labs) while 125,000 worked in government labs.[14] These numbers had advanced substantially from the postwar lows of the 1940s, but by 1963 they had plateaued.

Almost everybody who looked closely at these numbers deemed them inadequate. Federal research expenditures had grown 35–40 percent per year since Sputnik.[15] The rapidly growing space program was hiring so many engineers and physical scientists by 1963 that it could easily absorb the entire output of the nation's graduate programs each year.[16] Projections just two years into the future suggested shortfalls of engineers and scientists of all types, and the wave of immigrant scientific talent which had energized the American scientific effort in the 1950s had largely ended.[17] A PSAC committee warned in December 1962 of "impending shortages of talented, highly trained scientists and engineers" that would undermine "vital national commitments" and recommended growing the pipeline of doctorally trained engineers and physical scientists from 3,000 per year to 7,500 by the end of the decade.[18] In Wiesner's assessment, defense and space programs were adding "massive new demands" to the nation's scientific workforce, and a constant seepage of the most talented individuals from government to industry was hurting government research and degrading the government's long-term ability to carry on necessary research and development.[19] Moreover, nondefense research suffered secondary repercussions as talent flowed upward to the Pentagon, industrial labs, and the universities.

The country lost scientific potential at every level of the educational system, from the elementary grades to the postdoctoral programs. Few of the nation's high school students took advanced science classes and fewer still pursued science concentrations in college.[20] One stunning marker of the nation's inability to nurture scientific potential was the miserable benchmark of 4 percent – the portion of college engineering majors earning the BS degree who would go on to earn any sort of graduate degree.[21] And even those young men and women who completed graduate programs were unlikely to work in government labs, decamping for industry and universities as quickly as they could be trained. By contrast, the Soviet Union continuously marshaled its education and training efforts to produce a growing pool of industrial engineers employed in the nation's munitions and mechanical industries. Frederick Terman, the dean of Stanford's engineering school, and a caustic observer of the nation's scientific preparedness, noted the Soviet Union's sophisticated computers, radar sets, jet airplanes, and general electronics industry. He wrote in 1963,

> It has hydrogen and atomic bombs, and the largest bomb that has ever been exploded. . . . Russia's ballistic missiles for military and space have reached each new level of achievement both sooner and with bigger payloads than have our missiles. . . . The fact that we have better plumbing, more bathrooms per hundred people, electric toothbrushes, and more variety in women's hats may result in a civilization that we feel is superior, but these things don't count in the military and space competitions.[22]

Gaps in science education started at the high school level, where an explosion in postwar births was producing a shortage of high school teachers by the late 1950s. At the same time, stagnant teacher salaries were persuading many of the most gifted and talented young people with scientific training to eschew teaching and head to industry or applied scientific fields. With college enrollments slated to rise from 3 million to 4.5 million through the 1960s, and increasing demand for professionals with science training of all types, high schools were badly challenged in their faculty recruiting efforts. Industrial scientists, for example, could expect their salaries to double in their first five years out of college, while high school science teachers could only hope to maintain parity with inflation. Glenn Seaborg noted that "electricians, plumbers auto workers, railroad conductors, and engineers earn more than teachers," and that in constant dollars, teaching salaries had actually *declined* since the 1940s.[23]

As had long been true, the United States bled talented high school students to the workforce, luring only a small portion to college even after the enrollment expansions brought by the G. I. bill. This was partially due to broad demographic and cultural forces at work in the United States, where a substantial portion of the population had long been dismissive of the utility of higher education, but again the problem lay at least in part with the teaching force of the high schools. A few pilot programs in the 1960s – notably one between RCA and the New York City Board of Education – tried to match local industrial labs with school systems, but the project was dependent on physical proximity and was largely unscalable.[24] More promising was the hard, fundamental work of recruiting better, more inspiring, and more effective science teachers to the nation's high schools where they could excite young people and counsel them into college science majors. It was difficult to isolate the impact of a particularly effective teacher, but most scientists could point to a handful of influential mentors who had steered them into the discipline. One staff member on President Eisenhower's PSAC averred, "I am sure that many of these (college matriculants) would be much more strongly conditioned toward productive scientific careers if the average quality of high school science teaching were increased."[25]

Part of the problem was the nation's reluctance to invest in good teaching or to even acknowledge that the skill could be taught. The best high school and college science teachers had completed at least some graduate work in their fields, but graduate programs nearly entirely eschewed training in pedagogy. One survey conducted in the 1970s found that nearly 80 percent of all holders of graduate degrees in chemistry, physics, and biology had published work beyond their dissertation while fewer than 50 percent had received any sort of pedagogical training.[26] That is, the graduate programs focused their efforts on recruiting and training the next generation of researchers and assumed that pedagogy at any level would be picked up through osmosis. An internal White House missive described the "great teacher" as the "goose that lays the golden eggs" but admitted that few institutional norms supported it.[27]

Perhaps even more elusive was raising the level of scientific awareness in the general population – most of which remained ignorant of the most basic scientific advances in chemistry, physics, and biology over the previous five decades. Such an effort would require not better college teaching, or even better high school teaching, but perhaps better teaching in junior high schools, vocational tracks, and the burgeoning community college sector. A related challenge was delivering the required survey science courses for nonscience majors in the nation's four-year colleges. At every level, the nation's investment in scientific training and progress emphasized diverting funds to the most talented and advanced students who showed the most promise for innovative research while ignoring the general scientific ignorance of most of the population. The production pipeline simply failed to match the population's needs.

Wolfgang Panofsky, the director of Stanford's high-energy physics lab, concluded that even Stanford could not effectively teach basic science courses to the mass of its talented freshmen. He wrote in 1960,

> [I]n almost all universities teaching responsibilities of the faculty include elementary instruction which usually does not overlap with research. . . .

This is, in fact, a very serious problem, and I believe it is a very important reason why recently a number of experienced research people have left universities to work at institutions concerned with research only.[28]

On a pithier note, Jerrold Zacharias, a professor of physics at MIT and a strong advocate for educational reform, stated one of his major objectives for American science education: "The power of scientific reasoning and logic should be instilled in every educated person."[29]

The government lacked an effective response. The vast majority of government funding for science, whether through the DoD, the AEC, the NSF, or the NIH, was slated to support either research or doctoral training for research. While the NSF maintained a small "Science Faculty Fellows Program" for mid-career college science teachers, the program was minuscule compared to the billions pouring into the space and weapons programs and could support only 100 fellows each year.[30] A proposed National Education Improvement Act of 1963 to expand fellowship and loan programs failed to gain traction and produced little.[31]

An opposing camp, however, suggested that the real issue was not so much quantity but quality. Bachelor-trained engineers were starting at salaries no different than entry-level administrators, corporate salesmen, and radio technicians, and trailing in pay in later years, suggesting that they had emerged from their undergraduate programs with relatively pedestrian skills.[32] Most universities across the nation, including the nation's flagship state research institutions, admitted nearly all high school graduates who applied. Many of the nation's BS programs were undemanding and produced graduates who were unqualified for advanced analytical projects. Most graduates took line jobs in industry where their work was more technical than innovative.[33]

The shortage was at the elite end of the professions, where the United States seemed able to recruit and train only a handful of scientists and engineers capable of performing innovative and sophisticated research. George Kistiakowsky, the chairman of PSAC from 1959 to 1971 and a professor of physics at Harvard, admitted that within science "the excellent is not just better than the ordinary; it is almost all that matters," and that "first-rate work within the sciences is the only path to continued American economic and military supremacy."[34] Eugene Wigner, the Hungarian émigré physicist who would win a Nobel Prize later that decade, wrote in 1960 that while the United States maintained a thin edge over the Soviet Union in the pure sciences, it was now trailing in the applied sciences, where the Soviet Union had focused a substantial portion of its resources over the previous 15 years. "The [Soviet Union] strains every muscle to get ahead of the [United States]," he wrote. "Unless we can more nearly parallel their purposefulness, they shall prevail."[35]

Terman, of Stanford, as usual, was a source of most elitist rhetoric and harshest criticism, warning of the fallout from American intellectual mediocrity and lethargy and of the need to create more concentrated centers of excellence rather than broad scientific literacy and education. National scientific supremacy could shift quickly, from the French medical juggernaut of the early 19th century to Germany's superior chemical industries in the later 19th century, to the American and British wartime successes in radar and fission, to rising Soviet and Japanese accomplishments in the engineering fields. "Russia could easily be on top in one or two decades," Terman warned.[36] Experience in industrial labs hardly substituted for excellent classroom teaching and an apprenticeship in a university lab. "An inexperienced college graduate does not obtain a grasp of quantum electronics by on-the-job experience" he warned.[37]

The remedy, in Terman's reckoning, was not more teacher training, nor even more college students studying physics and engineering, but rather more intensive graduate training at the highest level for the smartest students who had already completed college. The nation must invest carefully in only the most selective and promising departments to double the number of universities capable of excellent work.[38] The tallest trees must be made taller still, and government resources should be carefully conserved for those programs capable of joining the top tier – the "centers of excellence" and "points of strength."[39]

Terman's response rested on government funding. No American university in 1960 had adequate endowment income to fund large teams of top-tier scientists for years on end, and the volume of philanthropic support was wholly inadequate to meet the demand.[40] The NSF's 1962 budget of $12.5 million could fund only a very small portion of the nation's total training needs at the advanced master's and doctoral levels. One estimate placed the total financial need for graduate training alone in the engineering and physical sciences at $150 million, rising to $300 million by the end of the decade.[41]

In addition to a nationwide shortfall of well-trained engineers and scientists, the government was hampered by its inability to retain scientific talent once it emerged from the graduate programs. Growing DoD and NASA research budgets actually exacerbated the problem, as industrial research firms, flush with money from their government contracts, used their inflated coffers to bid up the price of scientific talent. By the early 1960s, aerospace engineers holding the PhD commanded double the starting salary of other recently minted PhDs, and senior government scientists were taking early retirement to decamp to industry.[42] The government pay scale could not be adjusted quickly enough to accommodate conditions in the scientific labor market and lacked entirely a categorical distinction between PhD scientists emerging directly from graduate schools and the more valuable and productive PhD scientists coming from postdoctoral fellowships.[43]

The problem was caused by projecting civil service job categories (and wage scales) onto scientists. Talented scientists seemed to show a surprising awareness of their own market value, cognizant of their ability to garner superior outside pay packages and play one potential employer against another. Scientists were highly aware of their own ranking within the global catalogue of scientific talent and expected to be compensated in accordance with their eminence and accomplishments. The government pay scale, by contrast, tended to rigidly sort personnel along crude benchmarks such as highest-degree-earned or years-in-position, with little or no recognition of gradations in the stature of degree programs or productivity within rank.

The system produced a cohort of elite university graduates who eschewed government work. In 1961, for example, only 11 students in the MIT graduating class accepted federal employment, and all of these seven were from the bottom half of the class.[44] A 1961 report from the Federal Council for Science and Technology concluded, in lifeless bureaucratese, "The resulting compression of top government salaries in comparison with those for the lower grades is now a well-recognized structural misalignment."[45] At the time, scientists at the GS-13 job rating in government earned just over $11,000 per year and the most senior government scientists, at the GS-18 level, earned $18,500. By contrast, full professors in the scientific disciplines at the most competitive universities commanded salaries approaching $35,000. The Federal Council concluded, "The inability even to approach the level of compensation offered by private enterprise has tied the hands of government recruiters and executives who search for highly qualified and highly able people."[46] The work-around – private government-contracting research firms – was an imperfect solution, creating the odd (and unstable) pattern of government scientific administrators overseeing junior colleagues in the contracting workforce who earned double or triple their salary.[47] A concerned Harold Brown, the director of Defense Research and Engineering, admitted, "There are just too many easy avenues of departure by which a talented scientist can leave the Department of Defense."[48] President Kennedy singled out the problem in a strongly worded letter to Congress in 1962, writing,

> [T]he heads of the government's three largest research and development agencies – the Department of Defense, the Atomic Energy Commission, and the National Aeronautics and Space Administration – have stated in the strongest terms the need to raise federal salaries, especially in the higher grades, if the government is to obtain and hold first-class scientists, engineers, and administrators.[49]

If nurturing and retaining scientific talent was hard, developing diversity in the nation's scientific workforce was even harder. The profession was populated almost entirely by white men. Black Americans rarely applied to universities with well-funded research programs, and even more rarely pursued science and engineering at the graduate level, whether blocked by institutional racism, inadequate preparation in high school, lack of family or community support, or some combination thereof.[50] Even more glaring was the absence of women at the higher echelons of science – more perplexing insofar as female college attendance rates were approaching male rates by 1962, and graduate and professional schools all admitted female applicants, albeit at levels reflecting the substantially lower number of women in the applicant pool. President Eisenhower's Committee on Scientists and Engineers recommended in 1957 that the nation recruit more women into the scientific and engineering professions, support scientific preparation for women at young ages, and attack the institutional and cultural prejudice which prevented employers from seriously considering female job applicants.

In 1959, the American Council on Women in Science began a nationwide program of outreach to promising female high school and college science students. The effort was hardly revolutionary. Women had graduated from German universities in the scientific and medical disciplines in large numbers in the 1920s, and some of the earliest female scientists in the United States were refugees fleeing fascism and anti-Semitism in Europe. With the White House increasingly concerned about shortfalls in scientific talent, women collegians represented a large pool of untapped talent. The Department of Labor reported in 1960 of women "climbing the science ladder" and highlighted a female physicist working on cosmic ray effects and a female astronomer who had recently been named to chair a department at a major research university.[51] While tremendous social barriers (and institutional hostility) dissuaded many talented young women from pursuing scientific careers, the rewards were considerable. Women taking research and faculty positions in the natural sciences earned some of the highest salaries in the female workforce, and government funding programs did not officially discriminate on the basis of gender.[52]

Challenges in Basic Research

Although research budgets had grown exponentially across the funding agencies over a decade, funding for basic research continued to lag. In 1960, only about one-tenth of federal research expenditures went to support projects which could be classified as basic, and the amount of nonfederal support for basic research was trivial.[53] (In contrast, industry supported over a third of all research in the nation, but nearly all was for applied projects.)[54] Basic research faced barriers in winning funding or even in garnering public support. On an institutional basis, agencies were unlikely to support work which did not directly speak to their missions – nuclear arms for the AEC, ballistic missiles for the Army and the Navy, or radar and satellite tracking for the Air Force. Kistiakowsky noted a pattern of pushing basic research to "the other fellow" in the hope that fundamental breakthroughs funded by one agency would benefit all agencies.[55] Basic research was also hampered by its fragmented nature. Fundamental scientific problems often did not fall along clear disciplinary lines, and their value and promise were prone to subjective judgment. Alan Waterman, in explaining some of the early failures of the NSF to garner Congressional support, noted that government budgeting processes flowed from the principle that "each agency should only conduct basic research in support of its particular mission," and thus an independent funding agency which drew support from no one federal department was vulnerable to underfunding.[56] By 1960, the NSF could fund only one-third of submitted proposals deemed of fundable quality yet was unable to win additional funds from Congress.[57]

Americans' very pragmatism seemed to work against basic research which, in its ivory-towered loftiness, seemed inconsistent with the American *zeitgeist*. American scientists, whether inside the government or out, were asked to produce results and to show tangible advances flowing from those results. The sort of mathematical modeling and theoretical ideation which lay at the heart of innovative physics in the 1960s produced few tangible applications. Conrad Waddington, an eminent British biologist, paleontologist, and philosopher of science, noted upon visiting the United States that the American system of competitive grants, and the huge pressure on scientists to rapidly publish results, left little room for "contemplative, imaginative thought," resulting in a disproportionate share of the (admittedly impressive) American Nobel Prize haul to be for "discoveries on large machines . . . or the organization of large teams of research associates" rather than for fundamental breakthroughs.[58] Donald Hornig, a staff member on the OST, and later the special assistant to President Johnson, explained, "A system focused on the search for answers to current problems, could suffer from the absence of revolutionary new ideas not tightly linked to a recognized, stated goal."[59] Even Vice President Nixon, no fan of intellectual elitism, noted, "Very few of us have an adequate conception of the endless hours spent by scholars studying the electron, hours without which we would not have our television sets."[60]

Funding basic research was a problem because the NSF was problematic. Although its budget had grown over the decade, particularly since Sputnik, it was still responsible for less than 4 percent of all federal spending on research and development.[61] And because the foundation was disproportionately responsible for basic research, underfunding the foundation particularly hurt the nation's basic research enterprise. In 1962, even with a budget quintupled since the pre-Sputnik days, the NSF failed to fund over 600 research proposals deemed of "very high merit," along with 2,000 undergraduate and graduate fellowship applicants deemed highly qualified. Over 100 laboratory construction or expansion projects failed to be funded that year, condemning the losing universities to inadequate and obsolete facilities or, sometimes, to deferring research entirely. Computer purchases were being postponed; centrifuges unpurchased; accelerator time denied. Oceanographic research had ceased almost entirely as few reputable scientists could earn time on one of the handful of oceanographic research vessels afloat.[62]

The situation was slowly improving, however. Between 1958 and 1970, the NSF budget increased from $16 million to $200 million (a 12-fold expansion), even as the academic support budgets of the other major funding agencies grew at a quarter of that rate.[63] While the sheer size of the research budgets of the DoD, the AEC, and NASA continued to dwarf that of the NSF, the NSF was funding about 40 percent of all university-based research in the natural sciences by the end of the 1960s. Secretary of Defense Robert McNamara actually expressed concern, starting in the mid-1960s, of the diminishing role of the DoD in basic research, feeling that ceding the enterprise to the NSF would cost the DoD future breakthroughs in the scientific advances which underlay innovations in weapons systems.[64] Representative George Miller (D., CA), who chaired the House Committee on Science and Astronautics concurred, stating in 1968, "When the chips are down, we are not going to pull any applied rabbits from the scientific hat without knowing the basic tricks."[65]

By 1965, growth in the basic research budgets of the non-NSF agencies had declined from 26 percent per year to 14 percent per year, leading to a net slowdown in basic research growth, even as the NSF budget grew exponentially. NASA, in particular, had cut its university grant budget sharply after explosive growth in the first years after Sputnik while the DoD basic research stayed nearly flat through the latter half of the 1960s.[66] By 1967, the NSF's investment in basic (university-based) research equaled that of NASA and the AEC combined and was about 60 percent as large as that of the DoD.[67] Even so, government support of basic research still totaled only 5 percent of the total government research effort. The NSF was taking its rightful place in the basic research ferment, but basic research was still an underfunded stepchild of the government's research and development effort. Hornig testified in 1967 that Congress had not "fully grasped the stake which the nation has in a strong and well financed National Science Foundation."[68]

Applied research budgets, by contrast, remained large and robust. By 1966, the national laboratories alone were consuming $3 billion annually of AEC largesse, and other intramural government research programs – notably NASA, JPL, APL, and ONR – consumed nearly as much. These mission-focused laboratories, largely dedicated to weapons systems or the space program, were doing something closer to product development than true research; they resembled more closely industrial rather than university laboratories.[69] Alan Waterman, who had been replaced as NSF director in 1963 by Leland Haworth, found freedom in retirement to sharply criticize the nation's heavy emphasis on applied research, warning particularly of the growing portion of the nation's research budget consumed by the space program.[70]

At least a few people challenged the centrality of universities to basic research, which had been the centerpiece of the NSF's strategy from inception. Basic research, in theory, could be conducted in government or industrial labs, and certainly the federal government possessed the space and facilities to harbor it. President Kennedy's PSAC discussed moving at least some of the nation's basic research from university labs, and the DoD expressed willingness to dedicate some of its space to less applied work.[71]

Senior officials within the DoD, in fact, continually expressed concern over the department's underfunding of basic research through the 1960s, which remained about 2 percent of its research budget during that time.[72] Despite the hand-wringing, government labs simply seemed unwilling, or unable, to recruit the sort of superior and innovative scientists who could carry on fundamental research or to support them within an organizational structure which could accommodate unfettered and largely unsupervised creative work. The nation's basic research enterprise would continue to rely on a close relationship of federal funding agencies with the nation's two-dozen top research universities which, in turn, were getting more adept at exploiting the expanding stream of federal research grants.

The Government–University Axis

By 1963, the federal government was disbursing over $1 billion annually to universities to fund 17,000 graduate fellowships and nearly three-fourths of all sponsored research. Thirty percent of all engineering graduate students, along with 37 percent of physical science students, 46 percent of life science students, and 40 percent of students in the behavioral sciences were supported primarily by federal funds. About one-third of the funds came from the DoD and another third from the NIH, with the NSF providing 12 percent, NASA 8 percent, the AEC 6 percent, and the rest divided nearly equally between agriculture and commerce. This generous flow of funds substantially raised the scientific output of research universities but at the cost of possibly distorting institutional culture. Donald Hornig referenced the "healthy appetite for scholarship" generated by research universities but warned of the "scientific free enterprise system" which threatened to undermine institutional loyalty or even basic comity.[73]

Government-funded university laboratories were excellent incubators for the next generation of scientists but tended to undermine faculty commitment to teaching, generally, and to undergraduate teaching in particular. Faculty members hired under an implied or explicit expectation of regular outside funding sensibly shifted their energies toward garnering outside funds. While such a system expanded the ranks of universities in the United States capable of harboring first-tier scientific work, it at the same time undermined faculty commitment to institution building. As universities began to view star researchers as free agents, and the scientific faculty to be "bought and managed like baseball teams" (as stated in a scathing White House report), the willingness of the science faculty to invest in the university community, to teach large service courses, or even to take on the ordinary committee assignments expected of faculty, diminished.[74] "Soft-money" appointments, in which faculty were hired with no dedicated funding for the line, and contract bonuses, in which professors earned bonus pay based on garnering external grants, all contributed to declining university cohesion.[75]

External funding elevated graduate education at the expense of undergraduate education, which in turn allowed research universities to foist some of the more basic undergraduate teaching duties onto their graduate students. In theory, teaching obligations for graduate students were part of an apprenticeship to train future professors, but in practice the system tended to exploit graduate labor for some of the more mindless and tedious tasks – grading, running lab and review sections, troubleshooting student problems – while holding the more interesting tasks – designing courses and composing lectures – for the faculty. Faculty spoke of the "exploitation" of graduate students whose teaching was far more likely to benefit the university than the work of budding professor, and of the tendency for professors to withdraw from more teaching obligations than their grant proposals had defined.[76] Faculty who were heavily invested in research tended to emphasize their own research material in designing classes, distorting their syllabi and pulling departments to particularly narrow orientations. A White House panel warned in 1961 of the ways in which grant-driven remuneration could "contribute to a deterioration of the college community both financially and intellectually by emphasizing areas of immediate interest, rather than the broad base development of resources so essential to the educational needs of the students."[77]

An even more extreme situation existed at the medical schools, where faculty were so well funded that not only did *they* avoid teaching, but their doctoral students did too. Buoyed by a huge increase in NIH external grant funding, medical school teaching loads rapidly declined from two courses per semester to one course per semester to one course per year to three or four *lectures* per year. Faculty members had become, effectively, full-time research scientists. Such patterns affected the broader university community insofar as medical school faculty appointments became more attractive to ambitious young faculty in the arts and science departments, while graduate students in the basic sciences gradually migrated toward medical school labs where they could work on generous medical school grants with few teaching obligations. One professor of biology at Harvard rued the ways in which the medical school grants undermined his own department – drawing off talented teaching fellows and creating competition between the basic biological researcher harbored in his department and the more applied physiological and clinical researchers over at the medical school.[78]

One pervasive question was the issue of teaching–research balance. In theory, a professor with an ordinary teaching load of eight courses per year could benefit from buying out a quarter of her teaching to free herself to pursue research. Summer money added a solid three-month block of time for extensive focus and lab work, and more research funds could buy off further teaching obligations, allowing the professor to devote nearly two-thirds of her time around the calendar year to research. But at what point did the teaching obligation so decline as to fundamentally disassociate the work of new discovery with the work of assimilation, aggregation, and transmission? That is, if teaching and research were fundamentally compli-mentary activities, was there not a minimum amount of teaching in which a faculty member needed to engage to realize that synergy? Henry Bent, the dean of the graduate school at the University of Missouri, encouraged the federal funding agencies to strictly define support parameters in such a way as to force teaching obligations on grant recipients, create genuine teaching apprenticeships for graduate students, and generally guide universities to a more balanced faculty workload.[79]

The biggest concern was for the quality of undergraduate teaching. Logic dictated that faculty investing the bulk of their time and energy into research and graduate mentoring would underinvest in undergraduate ped-agogy. The physics community was concerned enough about this in the early 1960s to establish a Committee on Physics Faculty in Colleges, which surveyed college faculty and administrators around the country. The results, however, were surprising. While undergraduate students complained of the inaccessibility of faculty at research universities, they were no more likely to complain in those disciplines which received substantial external research funding (engineering and the hard sciences) than in those which did not (the arts and humanities). Moreover, in disciplines which were rapidly evolving (such as physics), research-active faculty were more likely to reconfigure course syllabi to include new discoveries, theories, and modes of thinking. That is, while the research-focused professors might not be investing in pedagogy and student interaction, they were investing their efforts in course design and content. Moreover, professors at small, less prestigious colleges, where the faculty had little research support, admitted to failing to modify course syllabi to reflect recent advances.[80]

Complicating the situation was the fact that students in the sciences, and particularly in the physical sciences and in engineering, were often the strongest and most academically dedicated students on campus. Arriving at college with better preparation and greater focus, these students not only could compensate for subpar pedagogy with increased dedication, but were more likely to seek out professors who brought innovation and cutting-edge insight into their classes. That is, undergraduate students who were interested in scientific careers were more likely than their peers to seek out professors who conducted research and who revised their own syllabi and teaching accordingly. A committee investigating these issues for the NIH in 1965 reported that "the availability of stimulating researcher-teachers, and the new areas of interest and inquiry which they bring to their instruction, arouse greater interest in students than did the exclusively clinical doctri-naire instruction which is steadily being supplanted."[81]

All in all, research on college teaching painted a complex and ambiguous picture. Funded faculty devoted less of their time to teaching but were more likely to inform their courses with current research. Such courses, while pos-sibly more poorly taught (from a point of traditional didacticism), nonetheless drew enrollment from some of the stronger and more ambitious students on campus who sought out courses taught by research-active faculty in which they would be exposed to more innovative material. And while the best graduate students did, indeed, avoid teaching by working for well-funded faculty who would not require them to teach, the strongest undergraduates tended to migrate to those universities which garnered more research funds and as such were better able to compensate for pedagogical weakness in the faculty. Increased research funding was forcing a broad sort among college students, in which the best graduate and undergraduate students flocked to universities winning the preponderance of federal research support. These institutions, now populated with more academically gifted students, could afford to invest less in traditional pedagogy and student support.

In a broader sense, research funding was both strengthening and weakening American universities. Increased funding had elevated at least a dozen universities into the top tier of research output over the previous 15 years. While in 1945 the country had fewer than ten universities capable of sustaining major research enterprises, that number had increased to 20 by 1960, and it would probably be 40 by 1975.[82] Strong students could now attend a greater selection of universities and still have a reasonable chance of being exposed to engaged, research-active faculty who imported some of the excitement and innovation of their labs into their classes. At the same time, American student bodies were becoming more tiered, with the most scholastically accomplished young men and women now choosing to attend a much smaller group of selective colleges. Whereas once highly talented college students were spread broadly throughout the nation's thousands of four-year colleges, increasingly the best students, particularly in the sciences, were choosing to attend a small subset of those schools.

The effect of federal funding on graduate and postdoctoral education was less ambiguous; government funding had greatly improved and expanded the quality and quantity of the programs. The country nearly doubled the number of PhDs awarded in the natural sciences in a decade (going from 3,800 in 1954 to 6,600 in 1964). At the same time, the ranks of the schools producing competitive PhDs had expanded, with the portion of all science PhDs produced in the top ten PhD factories declining from 46 percent to 35 percent each year. In most of the hard sciences, the portion of those earning a PhD who went on to a postdoctoral fellowship had doubled over a decade – from 15 percent in 1954 to 30 percent in 1964.[83]

Throughout, the PhD remained the pivotal degree in producing the next generation of researchers. None of the newly emerging alternative doctorates (such as the DPhil, EdD, PsyD, or DrPH) required the depth of research experience or the production of material of publishable quality. And even as more universities sought to enter the PhD-granting ranks, the degree tended to maintain its standard. In part this was a result of the professors, themselves, who did not wish to dilute the very degree which gave them their professional standing, and in part it flowed from the understanding that the degree not only include a research component, but that it actually be built around the research component. The need for rigorous research experience from the graduate students was part of the force impelling the faculty to pursue their own research; it would be a risible mentor who could not succeed in the very field in which she was guiding her protégé. A 1960 PSAC report on research and the universities repeatedly stressed the centrality of the research-intensive PhD program in promulgating and strengthening American science. The report condemned mediocre research as "worse than useless" and vigorously advocated for ever more rigorous PhD programs built around innovative research. "It is experience of research that makes a man a scientist," the report concluded.[84]

To an even greater degree than was true with undergraduate science students, graduate students in science tended to congregate at the best universities, in this case easily identified by those schools which year after year won the most external research support. With the strongest universities successfully recruiting the most productive researchers to their faculties, who in turn attracted the strongest graduate and even undergraduate students, a national phenomenon of "talent skimming" had begun.[85] The 50 schools which received 90 percent of the available research funds had a near lock on high-quality graduate students. Donald Hornig warned that these schools, if left unchecked, might "siphon off all the best talent from other institutions and colleges."[86] Others concurred. John Walsh, a science reporter and observer, described the evolving university ferment in 1964, with the strongest universities growing stronger still under the steady nourishment of federal funds. "These universities tend to pay the highest salaries, boast the most celebrated faculty members, attract the better undergraduate and graduate students, and award the most PhDs."[87]

Leland Haworth, the NSF's director, defended the phenomenon with a succinct, "Funds from all agencies go where the competence is."[88] But it was Haworth, in part, who was responsible for actually distorting the shape and distribution of that competence.[89] The federal government could, through a coordinated interagency effort, push funds out to lesser colleges and universities which might be able, in turn, to use those funds to recruit more talented faculty and students. Decisions by the AEC and ONR in the 1950s to place national laboratories more broadly around the country had demonstrated the ability of the federal government to create and distribute scientific talent and not merely to chase it where it lay. But such a policy would require reworking the system of project and investigator-driven grants which had proven so successful over the previous two decades. A 1965 report produced for the House Committee on Government Operations pleaded powerlessness:

> If smaller liberal arts colleges do not wish to do research, it is clear that federal research funds can have no direct impact. The only beneficial impact they have is to contribute to the total pool of PhDs, some of whom may be attracted to teaching in liberal arts colleges in the future.[90]

The conundrum for the government was that even as its funding further skewed the distribution of talent in American colleges, the effect of the funding on American science had been extraordinary. While American scientists had won just 17 percent of Nobel Prizes in the sciences between 1920 and 1940, they had won 53 percent of all such prizes since 1953. In other measures – peer-reviewed articles, international science awards, patents – American scientists were improving at similarly spectacular rates.[91] Identifying and supporting the most talented American scientists and scientists-in-training and aggregating them in a small and elite group of research universities produced terrific innovations and discoveries, even if it undermined the collegial culture of academe, or disenfranchised the millions of worthy-but-unextraordinary American students who could not earn a slot at one of the two-dozen research powerhouses. Haworth defended his policies by explaining that "a really good man, bright and well-educated, is far more important than several poorly-educated and marginally-creative individuals."[92]

The policy inadvertently affected the nation's less-privileged. While a few wealthy students in the past had traveled to attend a specific college, most Americans, historically, attended college within their own state and often within their own city or county. So long as thousands of small colleges could recruit reasonably competent faculty members, there was little or no professional cost to attending college close to home. But as national research universities skimmed off a greater share of the nation's research talent and attracted the majority of talented graduate students, undergraduate students who were disinclined or unable to travel for college were increasingly disadvantaged in their education and training. Centers of excellence produced award-winning research and successfully drew the nation's best graduate students to regional research centers. But these centers denuded the educational landscape for the thousands of small private colleges and regional masters-level universities on which the majority of American collegians depended. Concentrating the talent was terrific for those who could access the concentrated talent. For almost everyone else, it was a losing proposition.

Stanford's Tall Steeples

No university in the 1960s exploited these trends to establish its excellence as deftly as did Stanford. Heir to a solid engineering foundation laid down by the university's founders, the engineering school in the 1920s moved aggressively into graduate education, such that by 1928 a third of its faculty effort was in the graduate program.[93] While Stanford did not participate actively in developing tools and armaments for combat during World War II, and while it hardly ranked among the nation's outstanding engineering schools in 1945, it did have a solid reputation in the West, a good core graduate faculty, and a visionary dean in the person of Frederick Terman.

Terman had spent the war years working in Harvard's Radio Research Laboratory, a spin-off of MIT's Radiation Laboratory, focused on developing

electronic countermeasures to enemy radar. He returned to Stanford convinced that the future of American scientific and engineering innovations
would lie in government-sponsored research coupled with superlative graduate education, all anchored by faculty excellence. He developed a plan
for scholastic excellence based on sponsored research for both faculty and
graduate students – a faculty-first dogma in which the university would take
great pains to hire the most brilliant and innovative researchers of their generation and then give them great latitude in pursuing their personal research
projects. Terman envisioned an engineering school (and later a full university) which not only leveraged extramural grant funding to strengthen its
research, but which was actually constructed around extramural research
funding. In Terman's vision, competitive grant funding defined the work
of the university while teaching and university service flowed from faculty research interests. The university became a world–class research shop
in which talented and interested graduate students learned by doing, while
undergraduates progressed through proximity and osmosis. He dictated:

> Faculty quality is the *Sine Qua Non*. There is no substitute for faculty
> quality. While faculty committees may play checkers with the curricu-
> lum, and administrators may look for neglected areas, the reputation of
> an engineering school is dependent almost solely on the caliber of its
> faculty. Likewise, the soundest programs in sponsored research are those
> that are based upon the demonstrated productivity of the faculty, rather
> than upon the grantmanship of the administrators.[94]

All aspects of university processes flowed from this assumption. Department
chairs were appointed not for their skills in handling students or in developing curricula, but rather for their ability to recruit and hold the nation's most
promising researchers. New professors were expected to be both creative
and innovative researchers and have the ability to mentor a new generation
of researchers. (Classroom ability need only be "credible.") All researchers
needed to teach, and all teachers needed to conduct research. There could
be no division between teaching and research faculty, because for Terman
one could not credibly teach were one not actively engaged in research.

Terman adopted a philosophy of "tall steeples" – what would come to be
known elsewhere as centers of excellence:

> Academic prestige depends upon high but narrow steeples of academic
> excellence, rather than upon coverage of more modest height extending
> solidly over a broad discipline. Each steeple is formed by a small faculty
> group of experts in a narrow area of knowledge, and what counts is
> that the steeples be high for all to see and that they relate to something
> important. . . . The strategy thus indicated is to build up a very great
> faculty strength in a few important but narrow areas at the expense of
> broad coverage.[95]

The Stanford model proved to be perfectly matched to the enlarged flow of federal science and engineering funding to universities through the 1950s. A university producing the nation's best research in a handful of disciplines of great interest to the military (electrical engineering, aerospace engineering, nuclear engineering, theoretical physics, digital computing, solid state electronics) was perfectly poised to rise above its peers, attract the nation's strongest graduate students, and ultimately woo the most ambitious undergraduates as well. Faculty hired under this model could be counted on to bring in most, if not all, of their own (generous) salaries through grants, which could then be multiplied in indirects and lab salaries to spin off a team of talented doctoral and postgraduate students. Terman wrote,

> The view was that $10,000 per year from such [federal funding] sources would pay for the teaching time of a new faculty member, and if this man were properly selected he could obtain $50,000 per year for sponsored research, support a group of graduate students as research assistants, turn out one PhD per year, and push a Stanford steeple a little higher.[96]

The university benefited, too, from a stated political need for a West Coast center for applied research of interest to the AEC, the DoD, and later NASA. While both Caltech and Berkeley would share in the largesse diverted to the West (joined much later by the University of Washington), Stanford positioned itself most aggressively to exploit the growing trend.

Everything about the Stanford model projected elitism. In 1946, the faculty chose to offer the MS degree without a required thesis in an effort to focus its attention on doctoral students, who were capable of more sophisticated research. The students in the MS programs could thus be moved through quickly while drawing little faculty attention and thus free up faculty time to mentor the more substantive research projects of the doctoral and postdoctoral students. Output indicated success. In 1966–1967, the 133 engineering faculty members produced 497 MS graduates and 146 PhD graduates or 3.7 and 1.1 per faculty – a national record for graduate productivity.[97]

Stanford improved incrementally year by year in the two decades after World War II. In 1951, Stanford graduated 15 PhDs in engineering – ranking it about tenth in the country. NSF graduate support accelerated Stanford's rise, as large numbers of new graduate students chose to engage in training in the fertile milieu of California, lifting Stanford nationally to the second position (just behind MIT) in the number of NSF graduate fellows it was able to attract. The university received 2,200 applications to its graduate engineering programs in 1966, constituting 6 percent of all BS degrees in engineering conferred the year before.

The Terman model worked brilliantly for Stanford, but at a cost nationally for broader, softer, non-elite learning and maturation. It was a model particularly well adapted to the engineering disciplines, where youthful analytical brilliance tended to produce dazzling breakthroughs in practical problem-solving. Other disciplines, in the humanities and social sciences, as well as in law and management, which tended to draw more heavily on a scholar's insights into human behavior and organizational functioning, required greater seasoning, maturation, and, perhaps, wisdom. The model was irreplicable and unscalable, dependent as it was on talent skimming. While it was true that largesse in federal funding was increasing the number of research-intensive (later termed "R1") universities, these elite ranks could never number more than two dozen institutions, bound by the willingness of American taxpayers to fund university science and by the stated needs of the AEC, the DoD, and NASA to produce narrowly defined applied research. Stanford's gains were America's losses. A nation had only finite talent, which could be concentrated or dissipated. Concentrating it produced more celebrated science but a less enlightened, inspired, and informed populace.

Moreover, it was not clear that future engineering and scientific breakthroughs would come from small, faculty-centered labs. Radar, fission, guided missile systems, multistage rockets, the digital computer, and breakthroughs in particle physics had all been produced by large team efforts involving dozens or even hundreds of scientists and engineers. New work in astrophysics and theoretical physics presupposed access to enormously expensive pieces of equipment on which dozens of scientists might invest their efforts. The JPL, the APL, the national labs, and even Bell Labs were organized into teams of coordinated researchers each tackling a piece of a larger puzzle. At the dawn of the computer age, it would be sizable companies which would harness and organize the work of many coders, software developers, and hardware designers to produce functional computers which could be made available to large numbers of users.

The solution at Stanford and elsewhere was attracting a group of scientifically knowledgeable entrepreneurs to establish companies in physical proximity to the university in which faculty and recent graduates could work. The model was hardly new; General Electric, Monsanto, Bell Telephone, and the nation's pharmaceutical companies had long welcomed guest researchers from local universities to their premises as consultants or on a visiting basis. Ties between universities and industrial firms grew rapidly in the 1960s, particularly around Stanford in the southern half of the San Francisco peninsula (anchored by San Jose) and in the inner-ring suburbs of Boston along Route 128. These new technology companies drew talent from local universities while also providing opportunities for their faculty and graduates. At the same time, they served as economic drivers to attract young people to the area as likely to work in law and financial services as in research and development and business administration. Hornig described the growing phenomenon in 1967 as one of mutual benefit:

> While universities and research centers have much to offer local industrial progress, one also sees many cases where the pressures and demands

of industry are a major factor in the improvement of the university. . . . Industrial research and development needs to be tied to an industrial complex for it to be properly exploited.[98]

Stanford lay at the forefront of this phenomenon, which would later be replicated with varying success in New York, New Haven, Raleigh-Durham, Princeton, Albany, Los Angeles, and Austin. In 1951, Stanford established a 450-acre industrial park in a corner of its campus (later expanded to 700 acres) which came to host over three dozen firms together employing nearly 12,000 people, and by 1963, the university could count nearly 200 electronics firms clustered within 50 miles of its campus.[99] Such success was being replicated elsewhere, with Caltech taking credit for Beckman Instruments and Consolidated Electrodynamics; Bell Labs working closely with New York University; Rensselaer Polytechnic Institute teaming up with United Aircraft and General Electric (in conjunction with Cornell); and U.C. San Diego and Scripps working closely with Corvair. Cities around the country, wishing to emulate the economic growth of the San José region, established their own industrial parks and technology incubators – with Pittsburgh pairing with Carnegie Mellon, the University of Illinois, promoting two industrial parks in Champaign, and the University of Tennessee investing $2 million in the Arnold Engineering Center at Tullahoma.[100]

The effect, while nationwide, was particularly pronounced in California, where much of the nation's aerospace industry was located and thus benefiting from substantial research funds flowing from both the DoD and NASA. In 1965, California, Oregon, and Washington together constituted only 12 percent of the nation's population but garnered 44 percent of the government's research and development funding. The funding supercharged a Pacific-centric aerospace and defense industry, which was devouring nearly a third of all research and development funds (both government and corporate) and would continue to grow for another decade and a half. Such concentration further strengthened the position of Stanford, Caltech, and the UC Schools (and later Claremont McKenna and Harvey Mudd colleges) in being able to recruit talent which, in turn, earned more grant support, producing the next generation of talented young scientists and engineers – many of whom would be recruited locally and maintain their university ties.[101]

The industrial and high-technology incubators effectively emulated the relationship that medical schools maintained with teaching hospitals and that law schools created with legal services organizations, legal clinics, and the courts. Such university–industry relationships proved fruitful along multiple axes – in providing training and employment opportunities for current students and recent graduates; in providing consulting opportunities for faculty; in allowing industry to tap into academic talent; and in fostering mentoring relationships between senior industrial scientists and graduate and postdoctoral students. In almost every way possible, tighter relationships between government-funded university labs, graduate programs, and technology-based industry strengthened the work (and earning potential) of all involved.[102]

Shifting Budget Priorities

Through the 1960s, university administrators repeatedly questioned the system of channeling funds for research through individual investigators rather than through institutions. The system, as discussed, tended to concentrate talent and resources in a select few universities while actually weakening most others. While the system strengthened the scientific output of the nation, it balkanized universities and undermined governance. Universities could not plan effectively when each professor (particularly in the engineering, physical, and clinical sciences) was effectively a free agent, searching for outside support and buying out of his or her university commitments. One frustrated department chair at a large state university encouraged the White House to explore methods to convert some of the investigator grants to institutional grants in an effort to recreate an administrative structure more akin to the national laboratories, as opposed to the free-for-all evolving at the research universities. He wrote:

> If such a move toward funneling funds into proper channels by the government agencies is not developed, there is grave danger of . . . disrupting and weakening the institutions that they intend to help. . . . Lack of purposeful progress that can be achieved in basic research just as in applied research.[103]

But such protests found little footing in Washington. Rather, the current system of nurturing an elite cadre of research universities to the near exclusion of nearly all other institutions was producing the exact outcomes desired by the White House – steady parity with Soviet success in space, superiority in weaponry, and a string of scientific successes at the leading edge of particle physics, molecular biology, and the clinical sciences. Investigator-driven research might not be beneficial for the American university *system*, but it was highly beneficial to the select group of the nation's leading research universities as well as the faculty and students associated with them. A major planning report published in 1968 by the National Science Board (which maintained supervisory status over the NSF) encouraged more of the same: focusing the preponderance of research funds on a handful of research-intensive universities while encouraging most regional masters-level universities to focus on teaching. "No increase in the number of [doctoral] institutions is needed," stated the Board. Spreading the nation's finite talent across many universities would only weaken the nation's research effort. The Board warned liberal arts colleges away from trying to turn themselves into "second-rate universities" and generally cleaved to the tall steeple philosophy of higher education, in which a few national research powerhouses should be supported to the near exclusion of the majority of the nation's colleges and universities.[104]

Where money was to be spread around, it would continue to be through contracts rather than grants. The APL, the Aerospace Corporation, the System Development Corporation, MITRE, RAND, Lincoln Labs, the Ordnance Research Lab at Penn State, and the Institute for Defense Analysis continued to draw tens of millions of dollars of steady DoD contracts through the decade, allowing the government to effectively sidestep civil service hiring rules and pay above grade for elusive talent.[105] The AEC spread its largesse even more broadly, using the national laboratories (Argonne, Brookhaven, Oak Ridge, Los Alamos, and Livermore) for the bulk of its contract work on reactors and accelerators but expanding its footprint to installations managed by Iowa State (Ames), Sandia (Sandia Lab in New Mexico), DuPont (Savannah River), General Electric (Hanford, Knolls, and the Evendale nuclear propulsion facility), Pratt & Whitney (Connecticut Aircraft), Westinghouse (Bettis), the University of Rochester, and Princeton (Princeton-Pennsylvania Proton Accelerator). The AEC labs together kept 43,000 people on contract at budgets which exceeded $1 billion annually.

If there was any concern voiced about the system's effect on research universities, it was in adjusting the level of indirect support to ensure that universities were fairly compensated for the full burden which federal research grants placed on their infrastructure. President Kennedy took a personal interest in the issue, and in 1962 he called upon Congress to adjust the indirect rate upward to allow for administrative expenses and plant maintenance.[106] President Johnson reiterated the support in 1964, and Hornig warned that year that unreimbursed overhead was a "hidden form of cost sharing" requiring the university to divert resources (from the nonscience departments) to support work which ought to be fully funded by the taxpayer.[107] The concern was misleading. Little evidence suggested that the nation's research universities had suffered from the millions of dollars of federal grants won over the past two decades, and in fact most evidence pointed to the opposite: that the flow of grant money had strengthened the universities, enhanced their reputation, attracted more talented students at all levels of education, and produced more generous alumni.

If there was any shift in policy, it was toward broadening the NSF portfolio to include social sciences, applied research, and systems engineering. Military concerns drove this expansion, as both the Navy and the Army expressed the need for better empirical research on organizational design. Garrison Norton, the director of ONR in the mid-1960s, summarized the concern: "How can the Navy better organize itself for RDT&E [Research, Development, Technology, and Engineering]?"[108] Work within the contract labs often diverged from the ordnance, guidance, and nautical needs of the armed forces, and coordinating work within any specific lab posed frustrations for the career officers designed to oversee them. Competing research proposals flummoxed the operations heads of naval divisions while work between the three armed services was coordinated almost not at all. At the very least, ONR needed to create a systems engineering division to stream-line its organizational scheme and teach best practices to mid-level officers.

Civil unrest, riots, and frequent protests drew attention through the dec-ade to the very real needs of civil society.[109] Ensconced poverty in both White and Black areas, unequal access to educational and economic oppor-tunity, declining community institutions, and weakening family structure all argued for strengthening social science research. Jerome Wiesner, now back at MIT and serving as its provost, pushed the Johnson White House to expand the NSF's purview to social science with an understanding that social unrest demanded a more nuanced understanding of underlying social forces and dysfunction. He wrote:

> The social sciences are not going to point the way definitively to a solu-tion of all our ills, but they can give us a steadily deeper understanding of the working of our society, a better information base, and hence an enhanced ability to design our national programs to bring about desired objectives.[110]

Such proposals competed with the enormous funding needs of the space program which, at its height in 1966, consumed over 4 percent of the entire federal budget (and just under 1% of GDP).[111] In constant dollars, federal support for research and development actually declined in the late 1960s after growing exponentially over the previous decade. While total federal research support had nearly quadrupled between 1953 and 1966, it flattened out that year and thereupon declined by 13 percent over the next five years.[112]

Amid these competing demands, President Johnson was forced to make difficult funding decisions for research. In 1964, he denied funding for a proposed new accelerator in Madison, Wisconsin to be run jointly by the member schools of the Midwestern Universities Research Association (MURA).[113] In conjunction with Hornig, he laid out a series of science funding priorities which denied increased funding to particle physics, ques-tioning the need for accelerators with giga-volt capacities (beyond the exist-ing billion-volt machines) and sowing doubt on the relevance of continued work along these lines. "We cannot foresee at the present time the practical value of obtaining a clearer understanding of the nature of matter," explained Hornig.[114] The European consortium on nuclear research was building its own accelerators and storage rings at CERN in Geneva, and the United States was beginning to consider shifting some of the financial burden for particle research to its European counterparts. When asked by Congress in 1966 for the government's research priorities, Hornig's list represented, to a remarkable degree, military and space concerns, followed by advances in health care and agriculture, weather, transportation, and only last, general advances in scientific knowledge.[115]

One field of inquiry which the NSF did turn to in the 1960s was Antarctica.[116] The US Antarctic Research Program was a joint effort of 12 nations, which had signed the Antarctic Treaty in the wake of the International Geophysical Year, committing to reserving the continent for scientific research and wildlife preservation. Biologists, oceanographers, meteorologists, and glaciologists all worked together under federal grants to staff the existing McMurdo Station and to build a new Byrd Station several hundred miles inland.[117] Researchers paid particularly close attention to weather patterns (suspecting that Antarctic weather systems might play a disproportionate role in affecting global weather systems), the geomagnetic field of the area, and the movement of ice. Some suspected that the great West Antarctica Ice Sheet could be shrinking over time, possibly leading to global ocean rise and jeopardizing coastal cities.[118] The much larger East Antarctica Ice Sheet, 13,000 feet thick in places, contained as much as 80 percent of the Earth's fresh water but was stable.[119]

But dominating all was space. NASA had grown so rapidly that it appeared to be almost a government within a government. Funding requests appeared to be politically protected in a manner bewildering to skeptical members of Congress. The agency seemed immune to pressure from the DoD, the AEC, and even the White House. Its corps of astronauts was winning the adulation of the United States, and its televised missions drew some of the largest audiences in history. Its work would dominate American applied research through the end of the decade and into the 1970s, despite pleas for rebalancing research priorities. Compelling problems on Earth, even in the United States, beckoned, but all would have to compete with space for the next few years.

Notes

1 Jerome Wiesner, "Living with Science," delivered to the Federation of American Scientists, April 21, 1963, LH, 21: "Speeches by Others," p. 4.
2 The first modern semiconductor was produced at Bell Labs in Murray Hill, New Jersey in 1947 by a team led by William Shockley, which later garnered the Nobel Prize in physics. By 1954, Texas Instruments was producing digital calculators using 16 transistors. The first transistor-based radio was produced by Sony in 1957, using between four and eight transistors. Today, a modest laptop computer might have 3–4 billion transistors packed into its microprocessing chips.
3 See data in *Congressional Record*, 109:115, 7/30/63, pp. 1–2.
4 Ibid., p. 3.
5 In 1963, neither the Senate nor the House had a single doctorally trained scientist among their members.
6 "Congress Lacks Science Advisors," *Chemistry and Engineering News*, December 30, 1963, p. 23.
7 As late as 1960, David Beckler, the executive officer of the SAC, discussed the idea of a Department of Science with Eisenhower's senior staff but dismissed it as akin to "putting all government in a single department." Beckler to PSAC, July 13, 1960, AW, 32: PSAC, p. 8.
8 See Philip Handler, "Federal Science Policy: Roles of the President's Science Advisory Committee and the National Science Board," *Science*, 155:3766, March 3, 1967, pp. 1063–66.

9 See Jerome Wiesner, *Where Science and Politics Meet*, McGraw-Hill, 1965, pp. 45–49.

10 Eugene Skolnikoff, "Scientists and National Policy in the United States," July 23, 1962, DB, FA 965 16: Panel on International Science, p. 6.

11 Wiesner, *Where Science and Politics Meet*, p. 4.

12 Carl William Fischer, "Scientists and Statesmen," in Sanford Lakoff, ed., *Knowledge and Power: Essays on Science and Government*, The Free Press, 1966, pp. 325–26.

13 Quoted in Meg Greenfield, "Science Goes to Washington," *The Reporter*, September 26, 1963, p. 26.

14 Wiesner to JFK, May 8, 1963, JFK, POF, 85: OST, 8/2/63.

15 George Kistiakowsky, untitled internal briefing paper to the President, JFK, POF, 86: PSAC, 1/61–3/61, p. 2.

16 Frederick Terman, "Statement on Manpower Needs in Science and Technology," 1963, FT, series 10, box 2, p. 2.

17 Wiesner to JFK, May 8, 1963, JFK, POF, 85: OST, 8/2/63, p. 3.

18 PSAC, *Meeting Manpower Needs in Science and Technology*, GPO, December 12, 1962, pp. 1, 7.

19 Jerome Wiesner, untitled internal briefing paper to the President, JFK, POF, 86: PSAC, 1/61–3/61, p. 1.

20 Richard Nixon was particularly critical of the weakness in American high schools. He told one crowd that the United States needed to put "more fiber" into the high school curriculum and that "Soft subjects nurture flabby brains." From an Address to the National Nuclear Energy Congress, March 19, 1958, RG 255, NASA/Vanguard, 2: Press Releases, p. 4.

21 Frederick Terman, "Statement on Manpower Needs in Science and Technology," 1963, FT, series 10, box 2, p. 4.

22 Ibid., p. 7.

23 Glenn Seaborg, "The Future Through Science," *Science*, 124:3235, December 28, 1956, p. 1277.

24 See New York City Board of Education, "The David Sarnoff Industry-Science Teaching Program," in JFK, WHCF, 896: Science, 1962–63.

25 Willard to Kreidler, March 28, 1960, OSAST, 13: Basic Research, p. 3.

26 R.N. Kreidler, "The Proposal for a Non-Research Doctoral Degree," May 26, 1960, OSAST, 13: Basic Research.

27 Conrad to PSAC, "Strengthening of Instruction," March 31, 1960, OSAST, 13: Basic Research, p. 1.

28 Panofsky to Kreidler, July 19, 1960, OSAST, 13: Basic Research, p. 1.

29 Cited in minutes of PSAC, "Discussion on Scientific and Engineering Education," DDE, PSAC 5: Space, p. 2.

30 Harry Kelly, "NSF Activities in Scientific and Technical Personnel and Education," September 18, 1956, AW, GC: NSF, 1957, pp. 5–7.

31 Congress was more generous with veterans and children of wounded veterans, who received substantial educational support through the War Orphans Educational Assistant Act as well as an expansion of the National Defense Education Act. NSF funding gradually expanded into the fellowship programs but remained focused on research rather than on pedagogy.

32 Frederick Terman, "Observations on Engineering Education," *Aero/Space Engineering*, 17:12, December 1958, p. 48.

33 Ibid.

34 Panel on Basic Research and Graduate Education of the PSAC (George Kistiakowsky, chairman), "Scientific Progress, the Universities, and the Federal Government," 1960, OSAST, 13: Basic Research (6), p. 51.

35 Eugene Wigner, Untitled communication with the PSAC, "I deeply appreciate the opportunity to speak here today . . .," DDE, PSAC, 5: PSAC (2), p. 6.

36 Frederick Terman, "Engineering and Scientific Manpower for the Cold War: The Next Decade," FT, Series 10, box 1, p. 2.

37 Ibid., p. 3.

38 Kistiakowsky concurred. See *Scientific Progress, the Universities, and the Federal Government*, p. 52.

39 Frederick Terman, "Supplementary Remarks," FT, Series 10, box 1, p. 1.

40 A department faculty of 80 top-tier scientists and engineers, each drawing full compensation of some $30,000 per year (including fringe benefits) would have required endowment principal of $60 million to fund. At that time, Harvard, the nation's wealthiest university, held total endowment funds of approximately $200 million whose income was spread across nine schools and over a hundred departments.

41 Terman, "Engineering and Scientific Manpower for the Cold War: The Next Decade," pp. 5–6.

42 Wiesner to JFK, November 15, 1961, JFK, POF, 86: PSAC, 1/61–3/61, p. 3.

43 On this point, see Alan Waterman, "Notes on Problems Before the Panel on Basic Research and Graduate Education," June 2, 1962, OSAST, 13: Basic Research.

44 Harvey Brooks, untitled draft briefing paper, November 15, 1961, JFK, POF, 86: PSAC, 1/61–3/61, p. 8.

45 Federal Council for Science and Technology, "The Competition for Quality: Recommendations to Improve the Federal Government's Ability to Recruit and to Retain Superior Scientific Personnel," *IIR*, 46:3, April 14, 1961, p. 6.

46 Ibid., p. 10.

47 These firms, commonly called "Beltway Bandits," continue to perform a significant amount of government research and technical analysis.

48 Harold Brown, "Research and Engineering in the Defense Laboratories," *IIR*, 36:13, October 19, 1961, p. 6.

49 White House Press Release, April 30, 1962, AW 28: NSF Policy Papers.

50 Weisner wrote to Senator Hubert Humphrey (D., MN) in March 1962, in response to Humphrey's query about the absence of black students in graduate science programs.

> We rarely had an application [at MIT] from a Negro student, and I found on inquiry that this was true at most other good schools. . . . I am convinced that it is an extremely tough [problem] because it involves problems of elementary and secondary education, financial resources, professional motivation and more complicated psychological problems.
>
> From Wiesner to Humphrey, March 1, 1962, JFK,
> WHCF, 896: Sciences, 5/1/61–3/31/62

51 US Department of Labor, "Women are Climbing the Science Ladder," January 17, 1960, GPO, Washington, DC. Also, U.S. Department of Labor Women's Bureau, *Careers for Women in the Physical Sciences*, GPO, Washington, DC, 1959.

52 Officially. That is, no stated criteria for grant and fellowship programs indicated a preference for men. Members of funding committees, of course, brought their own biases and predispositions to the peer review process.

53 Clinton Anderson, "Scientific Advice for Congress," *Science*, 144:3614, April 3, 1964, pp. 29–31.

54 PSAC, "Support of Basic Research," July 12, 1960, OSAST, 1: Basic Research, 12/57–11/60, p. 2.

55 George Kistiakowsky, "Government and Science," March 9, 1960, DB, FA 965, 15: PSAC, p. 3.

56 Waterman was being paraphrased in a meeting of the Ad Hoc Panel on Basic Research of PSAC, September 21, 1960, DDE, PSAC, 3: Basic Research, p. 3.

57 Ibid., p. 4.

58 Quoted in Philip Boffey, "American Science Policy: OECD Publishes a Massive Critique," *Science*, 159:3811, January 12, 1968, p. 177.

59 Donald Hornig, Statement before a Subcommittee of the Senate Committee on Labor and Public Welfare on Amendments to the NSF Act of 1950," December 15, 1967, LBJ, DH, 7: Statements Before Congress, 1967, p. 4.

60 Quoted in H. M., "Nixon on Science: Policy Paper Emphasizes Basic Research; Call for Special Institutes," *Science*, 132:3429, September 16, 1960, p. 723.

61 Basic Charts and Tables Used by the National Science Foundation in appearing before the Subcommittee on Science, Research and Development of the Committee on Science and Astronautics, June 23, 1965, IIR, 36:7.

62 NSF Proposed Increased Budget Allowances for FY 1963, AW 28: NSF, Policy Papers.

63 Robert Geiger, "Universities and National Defense, 1945–1970," *Osiris*, 7, 1992, p. 44.

64 "Secretary McNamara on Research," *IIR*, 37:5, 1965.

65 Quoted in Donald Hornig, "Testimony before the Subcommittee on Science, Research and Development of the House Committee on Science and Astronautics," June 25, 1968, LBJ, DH, 7: Statements Before Congress, 1968, p. 5

66 C.E. Falk, "Possible Effects of the Declining Rate of Growth of Federal Funds for Academic R&D," June 1967, DB, FA 965, 40: NSF, Correspondence (1964–65).

67 "Testimony of Donald Hornig before the Senate Subcommittee of the Independent Offices Appropriations Bill for 1967," June 13, 1966, LBJ, DH, 7: Statements Before Congress, 1966, p. 2a.

68 Donald Hornig, "Statement before a Subcommittee of the Senate Committee on Labor and Welfare on Amendments to the NSF Act of 1950," December 15, 1967, LBJ, DH, 7: Statements Before Congress, 1967, pp. 1–2.

69 "Draft of Preliminary Report of the Panel on Government Laboratories," IIR, 46:6, 1965, p. 8.

70 David Tafler, "Scientist Warns Against Further Curtailment in Basic Research," *Montreal Gazette*, December 29, 1964.

71 See "Comments on Draft of Report of Panel on Basic Research and Graduate Education," 1960, OSAST, 13: Basic Research and Graduate Education.

72 See, for example, Beckler to Gray, July 15, 195, MIT, 195: SAC, "Basic Research Paper, 1957.

73 Donald Hornig, "Remarks to the Conference on Research Universities," October 8, 1964, LBJ, DH, 8: Addresses and Remarks 1964, pp. 1–4.

74 PSAC, *Scientific Progress, the Universities, and the Federal Government*, GPO, 1960, p. 20.

75 Ibid., p. 24.

76 Terman to Bent, August 10, 1960, OSAST, 13: Basic Research.

77 Colin MacLeod, et al., "The Federal Government and the Life Sciences," report to the PSAC, January 29–30, 1961, IIR, 46:3, p. 4.

78 Thimann to Kreidler, February 15, 1960, OSAST, 13: Basic Research.

79 Bent to Kreidler, July 29, 1960, OSAST, 13: Basic Research.

80 See "The Impact of Federal Research Programs on Higher Education," June 17, 1965, DB, FA 965, 15: PSAC, pp. 13–21.

81 Quoted in Ibid., p. 16.

82 The assessment is from *Scientific Progress, the Universities, and the Federal Government*, p. 14. I concur. In 2020, over 40 American universities each received over $50 million from the NSF alone. While the top universities still win a disproportionate share of funding (approaching $150 million), the funds are much more evenly distributed than they were a generation ago.

83 "The Impact of Federal Research Programs on Higher Education," June 17, 1965, DB, FA 965, 15: PSAC, pp. 7–10.

84 Apologies to women! From *Scientific Progress, the Universities, and the Federal Government*, p. 15.

85 It continues to this day. The strongest schools in the United States now draw huge applicant pools of extraordinarily qualified students. The most competitive now have admission rates of under 6 percent, and doctoral programs may be even more competitive still. Meanwhile, large numbers of less prestigious private colleges and universities face declining enrollments, shrinking budgets, and generally bleak futures.

86 Donald Hornig, "National Science Foundation," *C&EN*, July 5, 1965, p. 63.

87 John Walsh, "Centers of Excellence: New NSF Science Development Program Aims at 'Second 20' Universities," *Science*, 146:3651, December 18, 1964, p. 1563.

88 Leland Haworth, "Some Problem Areas in the Relationships Between Government and Universities," *National Academy of Sciences News Report*, 14:6, November–December 1964, p. 94.

89 While the NSF, in 1964, funded only about 40 percent of university research in the basic physical sciences, its growing prominence established its director as the spokesperson for federal extramural research policy.

90 "The Impact of Federal Research Programs on Higher Education," June 17, 1965, DB, FA 965, 15: PSAC, p. 24.

91 Haworth, "Some Problem Areas in the Relationships Between Government and Universities," p. 95.

92 Ibid.

93 The leader in this effort was the first dean of the Stanford School of Engineering, Theodore Hoover, older brother of President Herbert Hoover who had attended Stanford in its first class.

94 Frederick Terman, "The Development of an Engineering College Program," *Journal of Engineering Education*, 58:9, May 1968, p. 1054.

95 Ibid.

96 Ibid.

97 Ibid.

98 Donald Hornig, "Statement before a Subcommittee on Government Research, Senate Committee on Government Operations," July 11, 1967, LBJ DH, 7: Statements Before Congress, 1967, p. 6.

99 See Frederick Terman, "Stanford University," *Industrial Research*, April 1963, pp. 55–57.

100 See Terman, "The Newly Emerging Community of Technical Scholars," *Colorado and the New Technological Revolution, Proceedings of the University-Industry Liaison Conference*, April 1964, pp. 43–53.

101 Statistics are from Donald Hornig, "Statement before the Subcommittee on Employment and Manpower of the Senate Committee on Labor and Public Welfare," June 2, 1965, LJB, DH, 7: Statements Before Congress, 1965, pp. 2a, 6a.

102 Alvin Weinberg, the director of Oak Ridge National Laboratory in 1960, observed the untapped potential for industrial scientific expertise to mentor university students. See Weinberg to Seaborg, January 15, 1960, OSAST, 13: Basic Research.

103 McKew (?) to Kreidler, February 9, 1960, OSAST, 13: Basic Research, p. 2.

104 National Science Board, *Toward a Public Policy for Graduate Education in the Sciences*, GPO, Washington, DC, 1968, p. 33.

105 Wiesner to JFK, April 13, 1961, JFK, POF, 86: PSAC, 1/61–3/61.

106 White House Press Release, January 23, 1962, JFK, POF, 82: NSA, 1/61–3/61. Kennedy had been made aware of the issue while serving in Congress representing the Massachusetts 8th Congressional District, which included both MIT and Harvard.

107 Donald Hornig, "Statement before the Conference on Research Administration in Colleges and Universities," October 8, 1964, LBJ, DH, 8: Addresses and Remarks, 1964, p. 7.

108 Norton to Nitze, January 17, 1967, IIR, 37:7, p. 1.
109 In 1971, Alice Rivlin published *Systematic Thinking for Social Action* (Brookings), which spurred greater involvement by the academic social science community in the pressing social problems of the day.
110 Wiesner to Reuss, December 16, 1966, IIR, 45:7, p. 2.
111 In 1966, US government spending totaled $134 billion, while the NASA budget was just under $6 billion. Total GDP of the United States was $743 billion.
112 Figures from NSF, *National Patterns of Science and Technology Resources*, GPO, 1987, reprinted in Bruce Smith, *American Science Policy Since World War II*, Brookings Press, 1990, p. 80.
113 LBJ to Humphrey, January 16, 1964, LBJ, WHCF, AE, 1: 11/22/63–2/29/64.
114 Donald Hornig, "Statement before the Subcommittee on Research, Development, and Radiation, Joint Committee on Atomic Energy," March 5, 1965, LBJ, DH, 7: Statements Before Congress, 1965, p. 2.
115 Donald Hornig, "Statement before the Independent Offices of Appropriations Subcommittee, House Appropriations Committee," February 2, 1966, LJB, DH, 7: Statements Before Congress, 1966, p. 2.
116 Stans to Heads of Executive Departments, August 3, 1960, JFK, NSF, 282: NSF.
117 T.O. Jones, "Comments on the 1964–65 U.S. Antarctic Research Program," AW, 30: NSF, Antarctic Research Program.
118 Amazingly, scientists were already speculating about this possibility in 1965. See Ibid., p. 26.
119 United States Navy, "Welcome to Operation Deep Freeze," AW, 30: NSF, Antarctic Research Program.

11 The Space Race and National Prestige

Creating NASA

Sputnik had precipitated a scramble to match the Soviets with an American satellite. The failed Vanguard launches and rushed switch to the Atlas-based Explorer project suggested a thoughtlessness to the American space effort at the time – a series of face-saving reactions devoid of an organizing scheme. The surface chaos belied serious ongoing thinking about America's planned space program, however: the form it should take, its goals, its broad mission, and its organizational structure. Military and scientific planners had been considering space since at least the late 1940s, and even as the world concerned itself with the satellite race science, planners in the White House were laying out more careful and considered plans.

In late 1957, as the first two Vanguard efforts were very publicly failing, the NSF released a detailed plan proposing a national space effort independent of military objectives built on a new administrative infrastructure and possibly run through an entirely new agency. Questioning whether or not a nonmilitary space program was "essential to the national interest," and of what priority it should be given, the NSF considered such alternatives as a coordinated international effort (possibly including the Soviet Union and other Warsaw Pact nations), an *ad hoc* American program born of co-opted activities from existing federal agencies, and a more permanent American effort administered through a new space agency which would be capable of longer-term commitments.[1] Objectives might include research on the upper atmosphere, electromagnetic radiation, gravitational fields, astronomical phenomena, and biological processes in space. Practical applications might include better weather forecasting and facilitated global communication. The NSF urged that a space program of "exploration and research" be initiated "without delay."[2] Others concurred. Scientific consultants to PSAC envisioned a program geared initially to satellite-based observations of radiation, clouds, ionospheric activity, and geomagnetism, making use of orbiting telescopes, unmanned flights to the moon (and later Mars), and manned lunar landings.[3]

But military considerations could not be ignored. Sputnik had posed an amorphous threat to American scientific and engineering prowess, but it had

DOI: 10.4324/9781003363897-11

also posed a very real threat to American security because of a satellite's ability to take reconnaissance photographs. Space exploration had military implications, however idealistically it might be envisioned. In 1958, science advisors in the White House enumerated reasons to go into space, including exploration, defense, prestige, and scientific observation.[4] Vice President Richard Nixon gave voice to a more militarized vision of space when he deemed the space race "temporarily tied" with the launch of the United States' second successful satellite. The nation now seemed able to counter the fears produced by Sputnik: that it was weaker than the Soviet Union, that its scientists were inferior, and that its educational system was a failure. Rather, emphasized Nixon, the United States was "militarily stronger than any potential aggressor," its scientists were equal in quality to those of any other nation and, on balance, its education system was the equal of anywhere else.[5]

In launching a space program, President Eisenhower first needed to confront this challenge of separating the scientific and purely exploratory dimensions of space travel from its military applications. The challenge for the President was to create a space mission that went beyond military concerns and perhaps avoided military considerations entirely, even while acknowledging the military dimension of space. On the one hand, the military aspect could not be denied. Wallace Brode, the special assistant for scientific affairs at the State Department, stated succinctly: "The legitimate strong interest of the military in any space program is, of course, obvious."[6] On the other hand, space was too interesting and too vast to be seen as just another purview of military activity. An exasperated von Braun, in quibbling with Texas politicians over a future space program, commented that "space was bigger than Texas."[7]

While the armed services had proven themselves capable of funding and conducting excellent basic research, and of developing hardware for space flight and space exploration, the efforts were always subordinate to military considerations. Purely scientific efforts were likely to be marginalized by military leadership and subject to military budget constraints. Foreign governments might be less inclined to partner on space efforts if the American program was controlled by the military, and civilian scientists could be pressured to bias their research toward military application. One analyst on the Kennedy transition team noted pithily, "It is not to be expected that a military officer will easily give up an aircraft carrier or a fleet of bombers in order to get a satellite system."[8] In recruiting pure scientists to the effort, Bode summarized his insights: "This will surely happen most freely if it is obvious to the most casual glance that the program is really devoted to scientific advance in general and has not been subtly loaded in favor of purely military ends."[9]

Building a substantial nonmilitary space program raised the question of administrative structure: should an existing agency run the program, or did the government need to build a new agency from the ground up? Any program would need to be insulated from the military and possess contracting powers. A successful space program would require the development and purchase of expensive and sophisticated rockets, spacecraft, spacesuits, computing software, radio communications hardware, tracking facilities, and other components too numerous to foresee. The obvious candidates were the AEC, ARPA (DoD), and NACA, and while NACA had the word "aeronautics" in its title, it was not set up to award the sort of large multiyear contracts which a major space initiative would require. NACA, in fact, was really a collection of government laboratories staffed by permanent civil servant scientists with no record of developing or building its own rockets, spacecraft, or satellites. In 1958, it employed only about 8,000 people and had an annual budget of $80 million.[10] By contrast, the DoD had a long and successful record of contracting with private industrial and research and development firms for substantial projects, but always with a military bent. The AEC also contracted with civilian scientists and awarded grants and contracts to university-based scientists, but had a limited (nearly nonexistent) record in working with rocket technology. The AEC built bombs, but the Army, the Navy, and the Air Force delivered them.[11] The limitations of NACA, and the military mission embedded in virtually all DoD work, suggested to White House planners, as early as 1958, that a space program would require an entirely new agency reporting directly to the President.

In February 1958, at the peak of satellite fervor, President Eisenhower's science advisors recommended that he create a space program under civilian oversight, albeit with certain programs which were clearly military (such as missiles and antimissile defense) retained by ARPA and ONR. The need for civilian control required creating a new agency, born of NACA, but with substantial contracting authority which would report directly to the President. NACA's existing research programs would be reassigned to the new agency, along with certain defense rocketry programs such as Vanguard and perhaps the Army's Redstone program at Huntsville. The agency should have a presidentially appointed director and maintain contacts with the armed services through liaisons.[12]

Eisenhower responded with a speech to Congress on April 1 describing a proposed space program. Emphasizing the "high priority" of space for both science and national security, he recommended the creation of a new civilian agency – the National Aeronautics and Space Agency – overseen by a National Aeronautics and Space Board reporting directly to the President.[13] Congress responded quickly with the National Space Act, which the President signed it into law on July 29.[14] The new National Aeronautics and Space Administration (NASA) would have near monopoly on all matters related to space with the exception of developing missile-based delivery systems, orbiting weapons systems, and reconnaissance satellites, which would now fall to ARPA. The new agency's first-year budget, while relatively enormous compared to most federal agencies, was actually 20 percent less than the military's space budget of $294 million. Historian Audra Wolfe summed up the new agency's mission as "basic science, prestige, and a cover for military missions."[15]

NASA moved quickly. Less than a year after its creation it produced a ten-year plan, which included developing new space vehicles and scientific instruments. The existing rocket technology (primarily the Redstone, Atlas, and Vanguard rockets) was inadequate for lifting heavier payloads into high Earth orbit and much too weak for manned or unmanned lunar ventures. While the agency could adopt existing technology for its first mission – the manned Earth-orbiting Mercury program – anything beyond would require multistage rockets based on a planned Saturn design, whose first stage would be at least four times more powerful than existing Atlas rockets. An eventual manned lunar landing would require a complex rocket assembly many times more powerful than existing rockets; the first stage of the Saturn V which ultimately lifted astronauts to the moon generated nearly ten times the thrust of the Redstone rockets available at NASA's genesis.[16]

From its inception, NASA was both a scientific and engineering agency, with a dual mission of both collecting and analyzing data on space and also developing the hardware necessary to get into space. The development portion of the mission would be conducted nearly entirely with the work of contractors. The contracting portion would grow rapidly over the next half decade; by 1963, NASA employed more scientists and engineers than any other federal agency.[17] The agency became a major real estate holder when it purchased 80,000 acres of land adjacent to Cape Canaveral in Florida and a global presence as it rapidly constructed flight tracking stations around the world. In the speed and scope of its development it resembled nothing so closely as the Manhattan Project of 1944 – building enormous new installations rapidly while contracting with dozens of industrial firms. The Saturn assembly, on which all missions beyond the Mercury program would rely, was slated for initial testing as early as 1963.[18]

The Space Race

Despite Eisenhower's words to the contrary, many people in leadership positions in the Army's missile program promoted the idea of a race into space. General John Medaris, the head of the Huntsville operation, told a gathered crowd that the country was being obtuse if it refused to "make a race for it," and Keith Glennan, the first director of NASA urged Congress toward "a sense of urgency."[19] John Foster Dulles, the Secretary of State, emphasized the powerful propaganda value of being first into space, first with a manned spacecraft, and first to the moon. When the Soviets were first in any "spectacular" area, they gained as the United States fell behind. While there were few practical scientific payoffs from space travel, the psychological and diplomatic gains were considerable. Testifying before Congress, Dulles said,

> But the value from the point of view you put to me – you didn't put it to me in terms of the scientifics, but in terms of the psychological Cold War factor – and I was bound to say that I think they [the

Soviets] would make a great deal of use out of that, and would gain an advantage.[20]

Part of the sense of urgency was born of starting behind. Sputnik had shocked the country and signaled to space planners that the Soviets were farther ahead in rocket technology than had been thought. There was a reason for that. Since 1947, the United States had opted for smaller, more precise missiles rather than larger rocket boosters. A critical decision had come in 1954, when the Army opted for a 300,000 pound booster (the Atlas) rather than a recommended 600,000 pound booster, which would have allowed the country to maintain parity with the Russians. The Army had tailored its rocket technology to weapons delivery rather than space exploration, and thus by the late 1950s lacked the heavy machines needed to place probes on the moon or to put spacecraft past Venus or Mars. While this did not place the United States at a military disadvantage, it did undermine its technological bragging rights. Herbert York, the chief of research at ARPA, noted, "The matter of space . . . is more a matter of acute embarrassment that it is of national survival."[21] Given the starting discrepancies, the United States was likely to stay behind in booster technology through at least the mid-1960s.

Although level-headed analysts denied that the Soviets had a military advantage, some members of Congress and leaders in the executive branch responded to Sputnik impulsively. Representative David King (D. UT) called Soviet space superiority the "greatest potential threat to sovereignty which this country has ever faced" and called for a program to "put this country in the race and keep it there."[22] The usually staid Senate majority leader, Richard Russell (D., GA), warned his colleagues about the "death rays" that the Soviet Union was working toward.[23] Even George Kistiakowsky, Eisenhower's second science advisor, admitted that the optics and prestige associated with success in space justified investment in space travel and exploration beyond what scientific potential would justify. He wrote to John McCone, the chairman of the AEC, in 1960:

> In some instances, the issue of competition, that is of international prestige, is so important than an area (and I am thinking of outer space science and exploration as an example) must be given support somewhat in excess of what "cold scientific judgment" would allot it.[24]

By late 1959, the issue had become so heated and fraught that PSAC addressed it at length in a discussion paper titled "To Race or Not to Race?" PSAC analysts pointed out calmly that while the Soviets temporarily led in rocket technology, the United States would soon catch up. While the Soviets were exploiting aging rocket technology from a decade past, the United States either had, or would soon have, functioning models or prototypes of Scout, Juno II, Thor, and Atlas rockets and was developing the much larger Saturn. Current technology enabled the United States to boost heavy-enough payloads to place large satellites into high orbit, and the Saturn would soon supersede the Soviets' ability to place payloads into lunar orbit. Over the previous two years, the United States had successfully launched nine nonmilitary vehicles while the Soviets had launched only six. Nonetheless, a cold reckoning of 1959 capacity indicated that the Soviets were, at least temporarily, ahead. Their rockets were more reliable, accurate, and powerful, and the Soviets led not only in rocketry but in space science generally.[25]

One problem with engaging in a race was that speed tended to produce failures. The United States' first two satellite launch attempts had failed spectacularly in large part because they had pushed the Navy ahead of its scheduled timeline, and the many launch failures over the next two years could be traced to inadequate testing and preparation. Even the Saturn, on which the United States placed its hopes for future accomplishments, was not so much a technological innovation as an "extreme engineering elaboration" (in PSAC's analysis).[26] More thoughtful planning dictated greater investment in smaller and more reliable missile technology rather than in enormous space-bound rockets, yet the heated mood of the time compelled planners to make imprudent decisions. Even the Soviets recognized American sloppiness as being the root cause of failure; a Moscow missive concluded that the latest failure (an Atlas Able rocket) resulted from "the feverish speed with which the U.S. rocket space research program is being carried out."[27] Better would be to slow down, develop truly innovative rockets based on superior fuel and booster technology and more highly engineered parts, and take a longer, more strategic approach. PSAC wrote:

> Unfortunately, such a program means a delay of perhaps a year or two in the first flight tests of the initial big booster, because the present Saturn program may have to be re-oriented and that will take time. On the other hand, a truly useful new vehicle may be reached sooner and be better than the presently planned Saturn.[28]

Eisenhower, from the start, disavowed that the United States was in a race and consciously avoided using the term. To the contrary, his initial instinct was to invite the Soviet Union to join the United States in a program of cooperative, international space exploration. Under his leadership, the United States had proposed to the United Nations in January 1957 that all member nations pledge to commit themselves to exclusively peaceful and scientific space exploration, and in 1958 the President invited the Kremlin to join the United States in a combined space program. In a speech in March 1958, he took pains to explain that the plans for space exploration "reinforce my conviction that we and other nations have a great responsibility to promote the peaceful use of space and to utilize the new knowledge obtainable from space science and technology for the benefit of mankind."[29] Although the two nations traded barbs (the Soviets took pains to emphasize

American missile development while only tepidly endorsing a cooperative space program at the United Nations in 1958), Eisenhower persisted in inviting the Soviets to join the United States in at least limited areas of "scientific and peaceful uses of outer space" where the two nations might agree.[30] Such areas might include satellite tracking, astronomical observation, satellite-based communications, lunar probes, and an international space platform. Congress officially agreed, although individual members of Congress were skeptical. A major staff report to the Select Committee on Astronautics and Space Exploration dismissed international rivalry and emphasized that space exploration must be a "shared international concern" and that rival nations should refrain from "projecting current military rivalries to the limitless dimension of outer space."[31]

It soon became apparent, however, that the scientific and exploratory components of space travel were inseparable from establishing international prestige and projecting an image of technical and military superiority. Over the following seven years, the Soviet Union repeatedly succeeded in a series of space missions, which cemented its reputation for technical superiority in space technology and reinforced the idea that the Americans were lagging in space and in technological prowess generally. Between 1958 and 1960, the Soviets fired three successful probes toward the moon, including Lunik III, which passed behind the moon and transmitted the first images of its far side. In late 1960, the Soviets used their Venik rocket to launch three 1,900 pound probes toward Mars and Venus. (The White House estimated that the United States was still several years away from this feat.)[32] By 1963, they had launched a 3,100 pound probe toward the moon,[33] which was followed by a 1964 manned launch of a pilot, scientist, and physician, which orbited the Earth 16 times and descended successfully.[34] On September 3, 1965, the Soviet Union placed five satellites into the Earth's orbit, followed by six more on September 18.[35] When the Soviets reached their space zenith on October 18, 1967 with a "soft landing" on Venus (i.e., a controlled descent modulated through retro-rockets), James Webb, the director of NASA, could only acknowledge, if not actually celebrate, the achievement:

> The Soviet announcement that they have made a soft landing on the planet Venus represents an accomplishment any nation can be proud of. To go from Sputnik I to Venus IV in ten years illustrates the powerful base of technology being developed in the Soviet Union. The fact that this has been accomplished in connection with the fiftieth anniversary of the Communist Revolution is intended to encourage those in and out of the Soviet Union who believe the use of rocket technology to master and use the newly opened environment of space can become a major factor in the balance of technological power among nations.[36]

The Soviets logged failures as well, although their controlled media largely prevented the American public from learning of them. Many of their rockets exploded on the pad, and between 1959 and 1964 at least 20 launches toward the Moon, Venus, and Mars failed in one way or another.[37] Most horrific was a catastrophic incident on October 24, 1960, when an ICBM using a highly explosive fuel mixture misfired on its launch pad in Baikonur, Kazakhstan. The second stage ignited while the missile sat on the pad, instantly igniting the huge first-stage fuel tanks below. Dozens of technicians, scientists, military observers, and political supervisors were engulfed in the ensuing firestorm which may have reached 3,000 degrees. A perimeter fence, intended to prevent observers from getting too close to the launch site, instead trapped the men close to the inferno, preventing their escape. While the Soviets were highly secretive about the extent of the damage, later estimates put the loss of life at 130, including senior military observers and some of the nation's best rocket engineers.[38] The United States, by contrast, only officially lost three people through its entire program of manned launches.[39]

Kennedy, like Eisenhower, was reluctant to describe the US space program as a race. While there is little documentary accounting of his private conversations with staff and advisors, in his public statements and communications he refrained from using competitive language to describe the American space effort and always avoided calling it a race. In a letter to Congressman Overton Brooks (D., LA) of March 1961, the President explicitly emphasized the peaceful, civilian nature of the space program. He wrote:

It is not now, nor has it ever been, my intention to subordinate the activities in space of the National Aeronautics and Space Administration to those of the Department of Defense. I believe, as you do, that there are legitimate missions in space for which the military services should assume responsibility, but that there are major missions, such as the scientific unmanned and manned exploration of space and the application of space technology to the conduct of peaceful activities, which should be carried forward by our civilian space agency. I have been further assured by Dr. Weisner that it was not the intention of his space task force to recommend the restriction of the NASA [sic] to the area of scientific research in space. One of their strongest recommendations was, in fact, that vigorous leadership be provided by NASA in the area of non-military exploitation of space technology.[40]

Kennedy, in fact, claimed that he had always favored cooperation with the Soviets in space exploration and had even mentioned this priority in his inauguration speech. He and Khrushchev communicated periodically, and the President's letter to the Soviet premier of March 7, 1962, emphasized the "possibilities for substantive scientific and technical cooperation in manned and unmanned space investigations."[41] And while Khrushchev had evaded direct cooperative action, Kennedy asserted in a letter to Congressman Albert Thomas (D., TX) that he wished to work to "bring those barriers down."[42] "If cooperation is possible, we mean to cooperate." he wrote. "If cooperation is not possible – and as realists we must plan for this contingency too – then the same strong national effort will serve all free men's interest in space."[43] Others concurred. Congressman Victor Anfuso (D., NY) wrote in a passionate letter to Kennedy,

> While we must do everything we can to make our country and the world secure from possible future attacks from outer space, we must eventually combine the efforts of all nations on Earth to extract the full benefits from space exploration – the fulfillment of which would make wars among the world powers unnecessary.[44]

The race versus anti-race debate was more nuanced than our historical understanding might suggest. It was possible to see space as a scientific sphere of endeavor while still recognizing the prestige in getting there first. Through the Cold War, the Americans and Soviets, and indeed the entire Warsaw Pact and Western (NATO) alliance, competed through proxy endeavors, most notably at the Olympic Games but also through scientific prowess, vital health metrics, and even consumer appliances. During these years, athletic and intellectual achievements stood in as symbols of national prowess while medal counts, scientific publications, Nobel Prizes, and literary awards were followed closely for evidence of communist or capitalist dominance. During a celebrated visit to the American National Exhibition in Moscow in 1959, Vice President Nixon debated with Premier Khrushchev over the efficacy and usefulness of American consumer gadgets, such as a lemon squeezer, dishwasher, electric mixer, and toaster and boasted that such devices were available to middle-class American consumers and made their lives easier. (In a rejoinder, Khrushchev questioned whether the lemon squeezer was actually an improvement on squeezing lemons by hand.) Kennedy even admitted that the scientific motivations behind the American space program were probably competitive. When weighing the value of a Saturn program, he questioned whether "real success" could be achieved, or whether the United States was "so far behind now in this particular race we are going to be second in this decade."[45]

In denying the existence of a space race, Kennedy was continuing a strategy begun by Eisenhower in the aftermath of the Sputnik launch. The White House press team had taken pains to disavow a race, with the official posture summarized as, "We are not in any way racing with Russia; we are only racing with our schedule."[46] The NASA press office had gone as far as to issue a style guide for communications with the public, noting that government spokespeople should "avoid disparaging Soviet advances," refer explicitly to American "civilian scientists," and avoid emphasizing the "military implications of the Soviet effort."[47]

With Kennedy's assassination, however, and Johnson's ascendancy to the Presidency, the official message changed. Johnson was more comfortable than his predecessor in describing the space effort as a competition between rivals and of the need for urgency. In an interview with the *Houston Chronicle* in 1964, he explained that the United States should "never strive for a position less than first, and that this applied no less to a lunar landing than any other national endeavor."[48] As Vice President, Johnson had testified to the prestige associated with the "mounting number of our space successes."[49]

Johnson's stance was amplified by the administration's spokesperson on space issues, Edward Welsh, the executive secretary to the National Aeronautics and Space Council. In a speech before the Martin Company in 1964, Welsh stated strongly:

> It should not be possible for anyone to take seriously assertions that there is no danger to this country if it finds itself in second place in the space race. . . . Being second in space technology, in space competence, in space equipment, amounts to nothing less than endangering our freedom.[50]

And in a speech before Congress that same year, Welsh referred to the "serious competition which we dare not lose." He continued:

> In a philosophical sense, the most significant aspect of this national space program is that it disturbs the status quo. It makes obsolete those who think and act only in terms of the past. It inspires those who believe in the future of the country. It even puts stars in the eyes of those accustomed to judge progress solely by last year's profit and loss statements. The national space program is, in a real sense, a renaissance of the spirit of '76, the Declaration of Independence, and the Westward Movement.[51]

Through the decade, Welsh's rhetoric grew only more aggressive. Defending the NASA budget, he argued that being first on the moon indicated superiority in "military space activity" and that "dramatic" successes were the only politically meaningful signals in the absence of military action.[52] Success in space would translate into Johnson's continued political success while at the same time bolstering a pro-science attitude among the American electorate and even possibly appease American liberals, who might be uncomfortable with investment in weapons systems but could tolerate investment in space exploration.[53]

If indeed there was a space race, by 1966 the United States was pulling ahead. The Soviets had been first along most major lines of accomplishment including satellites, orbital flights, manned flight, lunar probes, lunar photos, and Venus and Mars missions. Now, however, the United States superseded those accomplishments with more sophisticated technology, more powerful boosters, and a more ambitious program. American photographs were clearer and American launches were more consistently successful. American rockets were larger and American tracking was now more reliable. The Mariner spacecraft had recently flown by Mars and returned data. American satellites were now hovering over most of the world in contrast to more limited Soviet coverage. And while both countries were collecting data, the Americans tended to publish more quickly and more comprehensively. A White House internal report concluded that year that, "an objective appraisal would put the U.S. clearly ahead now," although the Soviets might yet catch up in the future.[54]

Mission Development

The moon was the goal from the beginning. Although NASA created a diffuse mission for itself, which included unmanned missions to Venus and Mars and many unmanned probes and satellites with organic and inorganic payloads, it was rare to stray from the focus of landing a human being on the surface of the moon and bringing him (it would almost surely be a man) safely home again.[55] To that end, NASA's work through the 1960s followed a general trajectory of larger and more powerful boosters capable of lofting increasingly heavy loads into the skies and keeping them up there for longer periods of time. At the decade's dawn, NASA commenced Project Mercury – a series of seven manned launches atop Redstone and Atlas rockets. The first two, carrying Astronauts Alan Sheppard and Virgil ("Gus") Grissom simply went up to a height of about 120 miles, arced over the Atlantic, and came back down. The two flights lasted about 15 minutes each, produced a few minutes of weightlessness for the men, and were notable mostly in what did not happen – no catastrophic failures, no launch fires, no injuries to the astronauts. The most memorable aspect of the flights was the sinking of the Mercury II capsule when Grissom popped the hatch prematurely and the interior filled with water. Later flights included John Glenn's memorable three orbits around the Earth in the Mercury III mission, which drew tremendous attention and engendered national pride. His feat was doubled six months later by his colleague, Wally Schirra, who orbited six times, and Schirra, in turn, was superseded seven months later by Gordon Cooper, who orbited 22 times.[56]

Even as the Mercury flights were taking place, NASA committed itself to the two-man Gemini program, built upon the more powerful Titan II rocket with a spacecraft twice as heavy, aimed at orbital flights capable of staying up longer and docking with other craft while in orbit. The Gemini program explored the effects of longer space flights on biological systems and helped to overcome the challenge of controlling a multistage rocket – a critical skill in eventually flying to the moon. The Gemini program boosted ten two-man teams into orbit in 1965 and 1966, which stayed aloft for hundreds of hours rather than the 15 minutes of the early Mercury flights. By the time Gemini 7 landed, for example, its two astronauts, Frank Borman and James Lovell, had traveled more than five million miles.[57] Two astronauts, Ed White and Buzz Aldrin, successfully left the confines of the Gemini IV and Gemini XII capsules to "walk" in space, paving the way for the much longer extravehicular activities planned for the Apollo program.

The United States was not alone in these endeavors. The Soviets were matching the American achievements lockstep, putting up a three-man spacecraft in 1964, working outside the Voshkod II capsule while in orbit in 1965, launching an unmanned probe toward Mars, and putting up dozens of satellites. These achievements heightened pressure on NASA to move forward with lunar shots and ultimately a manned lunar landing.

Concurrent with the orbital and lunar missions, NASA commenced a program to launch unmanned spacecraft toward Venus and Mars which would send back photographs. These missions would be first steps toward landing unmanned probes on the surfaces of those planets (Mars first) in an effort to understand the chemistry and surface conditions of the planet and search for any signs of life. (Barring life, NASA scientists hoped to find evidence of water, which nearly all evolutionary biologists considered essential for life.) A flyby mission for Jupiter was also planned in an effort to map that planet's magnetosphere, and possibly a mission to Mercury as well. In 1965, the Mariner 4 successfully traveled 325 million miles to Mars to send back photographs. By 1967, NASA was planning an orbiting laboratory for launch in the mid-1970s (the eventual Skylab) and soft-landing missions to Mars in 1973 and to Venus in 1975.[58]

But all was prologue to the manned moon expedition. In 1961, NASA had begun planning a much larger rocket dubbed Saturn. Saturn was envisioned to be a multistage rocket (first two stages and ultimately three) with five enormous rocket engines at the base of its first stage, each capable of generating 1.5 times the thrust of the entire Atlas rocket. By 1964, the first Saturn flights were pushing large payloads (up to 39,000 pounds) into the Earth's orbit – a "weightlifting record" in the official NASA record log.[59] The larger Saturn IB was successfully launched the following year. Two years later, President Johnson proudly announced the "awesome sight" of the launch of the Saturn V, the largest rocket ever launched.[60] In 1968, Apollo 8 made it all the way to the Moon atop a Saturn V, orbited, and splashed down in the Pacific Ocean six days later. The following year, on July 20, American astronauts Neil Armstrong and Buzz Aldrin walked on the moon, marking the culmination of an extraordinary decade-long national effort.

The cost of the moon landings and affiliated projects had been immense. From a $360 million budget in 1961 (already more than double the NSF budget that year), NASA's budget quickly grew to $1.8 billion the next year, $3.7 billion in 1963, $5.3 billion in 1964, and just over $6 billion each year for the rest of the decade. When added to DoD space programs, and modest investments from NOA, the United States spent just under $8 billion annually on space exploration and development through the mid-1960s.[61] Hornig noted in 1965 that the cost of the space program exceeded that of all government nonmilitary scientific programs combined. He testified to Congress,

"It is bigger than health research, atomic energy development, university science, agricultural science, water resources research, weather prediction and meteorological research, and oceanographic activities put together."[62]

Much of this money had gone to aerospace contractors. The Apollo program used over 20,000 contractors and subcontractors to develop and build the Saturn V, the Apollo spacecraft, communications equipment, food, protective gear, new materials suitable for spaceflight, the launchpad complex at Cape Canaveral, and the large new Mission Control facilities in Houston. While NASA tried to distribute contracts to universities and firms around the country (with particularly attention to the Southern states), members of Congress and well-connected party leaders pressured the space agency to direct funds to favored local firms. MIT, as usual, did well; its instrumentation lab (later Draper Instruments) was a major developer of the Apollo guidance system.[63] Others were less successful. Bosch Arma bid unsuccessfully for the Titan III guidance system, even after heavy lobbying from the New York congressional delegation, while a prominent Princeton architectural firm lobbied unsuccessfully for design contracts for the space tracking and control facilities, which were erected rapidly in the early 1960s.[64] Senator Dennis Chavez (D., CA) pushed for Lunar Module testing at White Sands, Carl Albert (D., OK) wanted the Lunar Module contract to go to Martin-Marietta (Grumman Aircraft won out), and Harrison Williams (D., NJ) made a plea for ITT.[65] All failed.

Political Squabbling

With such staggering budgets, extraordinary made-for-television visuals, and stirring imagery, space travel was highly politicized from the onset. John F. Kennedy tried to leverage the issue during the 1960 presidential race, blaming the American space lag on Eisenhower-era foot dragging. His Republican opponent, Richard Nixon, suggested that the real blame ought to be assigned to the (Democratic) Truman administration, claiming, "We entered the space competition some paces behind. We paid a penalty because the Truman administration discarded and ignored the implications of the long-range rocket." It was Eisenhower, claimed Nixon, who had accelerated missile and rocket development, closed the gap with the Soviets, and leveraged the brainpower of the German rocket scientists in Huntsville.[66] Upon becoming President, Kennedy enthusiastically endorsed Apollo, rightly understanding that a moon mission could excite the public, unite the nation, and demoralize the Soviets. James Webb, the NASA administrator, took advantage of the support to elicit a White House commitment for even greater funding for the Apollo program – $50 million in 1962 alone. The White House hardly balked.[67]

The moon mission garnered general support from members of the public who feared the "Red Menace" and staked American greatness on technological sophistication, innovation, and daring. At the same time, the White

House attempted to win over skeptics with more practical arguments based on job creation, improved consumer products, and general economic development. By 1963, NASA was employing 9,500 scientists and engineers in its own labs, 5,000 in universities, and 30,000 more in industrial plants.[68] The nation planned to invest over $10 billion in the rockets, spacecraft, and ground control infrastructure to reach the moon. More than 5,000 commercial firms were contracting with NASA for a huge variety of tasks, services, and products. One White House staffer justified the effort as investment in the American soul. "There has to be progress and creativity to insure a continuing dynamic character," he wrote. "That is what our space program can supply."[69]

But some Americans balked at the cost and general impulsiveness of the venture. Vannevar Bush, retired but still alert, explained to Webb that the moonshot was simply more expensive than the country could afford: "Its results, while interesting, are secondary to our national welfare," he wrote.[70] Concerns over national prestige had displaced legitimate scientific inquiry, and dreams of glory were undermining ordinary progress in science and engineering. "I hear that the program will be justified by its by-products," he wrote. "We might get a billion dollars' worth of benefit that way. I doubt if it would exceed this."[71] The space program fundamentally weakened science, in Bush's estimation; it misled the public "as to what science is," and it indulged a more "juvenile quest for the spectacular."[72] Bush exhorted Congress to cut the budget before more money could be wasted.

Sage minds agreed. David Bell, the federal budget director, questioned the wisdom of "staking so much emphasis and money on prestige," and encouraged the White House to seek out alternative, cheaper missions to advance scientific understanding.[73] Wiesner, while visibly behind the program, privately questioned the economic benefits and even suggested that the massive investment was diverting funds from potentially more productive science or engineering projects. The massive pull of intellectual resources into the space program was impoverishing other worthy research efforts.[74] Even the nation's children (no doubt encouraged by their parsimonious parents) questioned the wisdom of the effort. Thirteen-year-old Mary Lou Reitler wrote to the President, explaining:

> When God created the world, He sent man out to make a living with the tools He provided them with. They had to make their living on their own with what little they had. If He had wanted us to orbit the Earth, reach the moon, or live on any of the planets, I believe He would have put us up there Himself. . . . While our country is spending *billions* of dollars on things we can get along without, many refugees and other people are starving while trying to make a decent living to support their families. I think it is all just a waste of time and money.[75]

Ten-year old Kris Lendahl followed in the same vein: "Why does my Daddy have to pay taxes with six children?"[76]

Myer Feldman, the President's domestic policy advisor, responded at length to the young Mary Lou:

> A significant feature of our society is the right of each individual to determine the nature of God's intent in accordance with his own conscience. I would not, therefore, presume to suggest how you should resolve the issue you pose in your letter. . . . History is replete with examples of man pursuing knowledge with no expectation of any practical use, which later serve as the basis for developments making significant contributions to mankind. . . . [John Glenn] in part said, "But exploration and the pursuit of knowledge have always paid dividends in the long run – usually far greater than anything expected at the outset. . . . Knowledge begets knowledge. The more I see, the more impressed I am – not with how much we know – but with how tremendous the areas are that are as yet unexplored."[77]

More specifically, skeptics suggested that the space program was eroding support for investing in other research and development efforts or actually crowding out other scientific work. By late 1962, two-thirds of all government research funding (both basic and applied) was going toward defense, space, or nuclear technology, and none of these three areas was producing significant civilian innovation.[78] In space, particularly, new products and devices tended to be highly tailored toward the unique requirements of space travel; they produced few civilian spin-offs while drawing a disproportionate share of the nation's best science and engineering talent. A special committee on technology and the economy chaired by Walter Heller, the chairman of the President's Council of Economic Advisors, concluded, "This situation is increasingly aggravated as more and more of our best technological brains – our most critical resource – are exclusively devoted to space and defense work."[79] Work on process innovation, consumer goods, manufacturing processes, data analysis, digital storage, and innovation in the energy sector was being underfunded. The government was skimming off so much research and engineering talent that two-thirds of all trained engineers in the nation by the mid-1960s were working either directly for the government or for a government contractor on a government program. Philip Abelson, the director of the Carnegie Institute's Geophysical Laboratory, stated succinctly, "We have a limited pool of genius."[80] One unfortunate result was the flat level of annual patent filings since 1940, despite a threefold growth in the number of graduate-trained scientists and engineers being produced yearly.[81]

Agencies in other sectors criticized the space program for the inordinate public resources it drew from competing programs.[82] Foundation executives, for example, criticized the diversion of funds away from the very pressing Earth-bound problems to the fantastical world of space travel. Warren Weaver, a vice president at the Sloan-Kettering Institute for Cancer

Research, noted that the cost of Apollo could be used to give every teacher in the country a 10 percent raise over ten years, with money left over to educate 50,000 new scientists and engineers, create ten new medical schools, and create three more Rockefeller Foundations.[83]

It was difficult to know who was right. Proponents were correct in that scientific insights often produced unexpected applications and that scientific research tended to produce economic growth, albeit in a nonlinear way. But skeptics were right, too, in that the Apollo effort maintained an extraordinarily narrow focus, tailored as it was to the unique demands of landing a spacecraft on the moon and safely returning its crew. Space posed unique engineering challenges – extremes temperatures, a total vacuum, the need to escape the Earth's gravity, unique guidance demands, the biological dangers of a deoxygenated environment, and extreme gravitational forces. It was not obvious how solutions to these many challenges would be applicable to more pedestrian problems. The mere need of a rocket engine to carry its own oxygen, for example, made its design irrelevant to ordinary airplanes which flew in the oxygen-rich atmosphere. Many of the engineering innovations required for space flight – carrying and pumping hundreds of thousands of gallons of supercooled, highly pressurized liquid oxygen among them – were simply immaterial to ordinary flight and propulsion.

Moreover, Apollo was not even that innovative. The Saturn rocket design was nothing more than a very large Redstone rocket, and the enormous thrust produced at liftoff was created by clustering five F-1 rocket engines, themselves little advanced from the V-2 engines produced by the Germans in 1945. While the lunar module for the Apollo mission was produced wholly new, as was the command module heat shield, much of the mission was more an update and expansion rather than true innovation. Lee DuBridge, by then the president of the California Institute of Technology, noted with some distance,

> Yet many of us feel that the prestige and competitive factors have forced us to move too far too fast, to spend too much money and devote to much effort to the "spectacular" as contrasted to the purely scientific ventures.[84]

By 1964, President Johnson was beginning to question the cost of the space program. With his energy and focus pulled toward civil unrest and the growing conflict in Vietnam, he questioned the enormous portion of his discretionary budget going toward space. Advisors suggested to him that the whole effort be turned over to the United Nations or to an international space consortium which could spread the cost among many nations while de-escalating a race for prestige that appeared to be senseless. Webb disagreed stridently, arguing that being first on the moon would indicate leadership in space capability (for military as well as civilian purposes), that the effort would cement Johnson's reputation as a leader, and that success would bind together a nation riven by conflicts of war and race. "It is only dramatic successes that are politically meaningful," Webb argued, "especially where military achievements cannot be widely discussed and spin-offs to industry cannot be easily identified."[85] Perhaps anti-poverty programs could be coupled to the space mission, suggested Webb. NASA scientists could spend 10–50 percent of their time mentoring disadvantaged youth who were now largely discouraged from pursuing science.[86]

By this time, too, members of Congress were growing tired of the endless rhetoric invoking Soviet superiority and the red threat from the skies. None of the uniformed services prioritized space. In Vietnam, the barriers to American victory were of the most Earth-bound sort: jungle warfare, guerilla tactics, home-turf advantages, and demoralized troops. Racing to the moon promised no relief for the United States' stalled efforts in southeast Asia, just as it offered no relief for the impoverished, the racially oppressed, the sick, and the uneducated. Increasingly, Americans across the political spectrum questioned the wisdom of the effort. The Critical Issues Council of the Republican Citizens Committee chaired by Milton Eisenhower, the president of Johns Hopkins University, declared in 1964,

> We find no reason to believe that putting a man on the moon could contribute to our military strength. We see no evidence that urgent defense objectives warrant a crash program or that a meeting of deadlines such as 1970 serves an significant national objective.[87]

Senator William Fulbright (D., WI), always on the prowl for government waste, proposed cutting $267 million from the 1965 NASA budget and perhaps giving up on a lunar landing entirely.[88] Even Donald Hornig, now serving as the director of OST, questioned the huge investment of time, funds, and scientific resources to produce a mission which, if successful, would culminate in astronauts spending only a few hours on the surface of the moon and whose effective "working time" would probably be shorter still. Real insight would come only from longer lunar visits, drilling into the crust, and more sophisticated measurements – none slated to be part of the work of Apollo.[89] It was left to NASA's most pure booster, James Webb, to repeatedly assert fears of Soviet dominance and American decline, reminding the President that the Soviets would be far ahead of the Americans but for continued American investment in the space program. "At present budgetary levels," he wrote to the President in 1966, "the gap between their large booster capability and our own would not narrow and might widen."[90]

When Cold War boosterism failed to inspire support for the moon mission, Webb fell back on the immensity of the program, hoping that the sheer scale would speak for itself. By 1967, NASA was distributing funds broadly through the American manufacturing sector: to Boeing for the Saturn V first stage, to North American for the second, to Douglas for the third, to IBM for instrumentation, to Grumman for the lunar module, to North

American (again) for the command and service modules, and to General Electric for internal electronics. The smallest of these contracts ran to hundreds of millions of dollars; the largest to billions. NASA was either directly or indirectly employing 400,000 American workers and was approaching a total investment of $25 billion in its quest for the moon. "We have utilized the American industrial system flexibly and in ways that have added vast new strengths that have permeated practically every segment of our national economy," Webb wrote to Senator Clinton Anderson (D., NM) in 1967.[91] While this did not prove that the Soviets were either leading or trailing in the race to the moon, nor whether the race bore any significance for economic development or military parity, the numbing catalogue of statistics seemed persuasive in and of itself.

Making Sense of the Effort

The stirring news footage of the Apollo launches at Cape Kennedy (recently renamed from Cape Canaveral) coupled with remarkable images transmitted from the various missions themselves – the famous "Earthrise" photo taken from Apollo 8 and the stirring photos of Armstrong and Aldrin walking on the moon's surface from Apollo 11 – instilled pride in Americans. The Saturn V rocket bore no corporate logos or titles but rather was stamped simply "United States of America." The flag planted on the moon's surface was the Stars and Stripes; the plaque on the base of the lunar module was signed by the President of the United States. The achievement was a national one, to which all Americans had contributed through their taxes, votes, and goodwill. Curmudgeons and skeptics stayed quiet, at least for the week.

But what had really been accomplished and at what cost? Was the endeavor simply an exercise in national muscle flexing – a potent symbol of the nation's technological strength and resolve? Or were there legitimate scientific goals underlying what was largely a triumph of chemical, aerospace, and materials engineering, administrative efficacy, and unprecedented investment of public funds? Politicians, board members, and administrators could each draw lessons to support their view, whether James Webb who argued repeatedly that the Apollo landings were really just a first step in lengthier and more scientifically intense space expeditions, or skeptical scientists who pointed out that all of the data generated could have been collected by probes, telescopes, and unmanned flyby missions. Brooks Overton emphasized the need to demonstrate clear economic and civilian technological payoffs from the effort – better communications and weather forecasting, for example – while some analysts suggested that tangible applications were already being realized in the form of improved aircraft wings, high-thrust engines, vertical takeoff military aircraft, and solar cells.[92]

Some scientists suggested that the lunar landing had not produced important scientific research because it was not designed to do so. Rather, it was a first step toward subsequent, longer missions which would include lunar rovers, drilling into the lunar crust, building lunar bases, taking mineral samples from a wide range of locations, and many more observations of the effects of space travel and weightlessness on the human body and other organic systems. It was too early to pass judgment because space had hardly begun to yield it secrets. Donald Hornig testified,

> We cannot now say, with any sensible certainty, whether doing so would be feasible or worthwhile. . . . I cannot now say whether it would be more in the national interest to spend the large amounts required on lunar bases.[93]

Arguing against further manned spaceflight was the fact that it was hugely difficult, predicated as it was on sustaining life in a total vacuum under extreme conditions of temperature and gravitational force. Acceleration on the way up imposed a 7G force on the men; reentry heated the outside of the capsule to 600 degrees. Human crew members required exercise, hydration, mental stimulation, and near-perfect protection from the elements. Any misfunction could bring the mission to a catastrophic and very public end. Hornig offered that the manned space flight was so costly as to exceed its value at present. "We must make major technological efforts to find ways and means of making maned systems less complex, more reliable, and much less expensive."[94]

Arguing against a pause were those who valued momentum. The enormous NASA complex of laboratories, engineering workshops, launch and control facilities, and contracting relationships could not be quickly restarted once mothballed. The program had huge support and momentum in the wake of the lunar landing, but that support might prove fickle once the flow of images from the moon and the nearby planets ceased. NASA was a creature of politics no less than any government program and as such drew its ultimate support from the preference of the voters. DuBridge expressed fear that momentum and ability might be lost should the United States not "push on to the planets."[95] Landing instruments on Mars and Venus, sending observation craft to Saturn and Jupiter, and putting a telescope or manned laboratory into orbit all relied on expanded rather than diminished commitment. Yet, by 1969, the NASA budget was already declining from its peak of $8 billion in 1965 to under $4 billion. Webb complained to the President that at that level, continued unmanned missions to Mars and Jupiter were probably impossible. The agency was laying off thousands each month, and lunar missions beyond 1972 were unlikely.[96]

The great dream of the space program's most admiring adherents – a manned voyage to Mars – looked increasingly unlikely. The full mission would take one to two years (compared to the two-week duration of the longest Apollo mission) and cost probably five times that of the moon mission.[97] More likely was a continuation of the Mariner probes to Venus and Mars (Mariner V passed within 2,000 miles of Venus in 1968) and the

Voyager program, which could potentially land a roving probe on Mars. One internal White House report asserted the need to demonstrate that the moon race was not "an end in itself."[98]

But was it? From its beginning, the US space program had superimposed national prestige and international rivalry on compelling engineering and organizational challenges. Science always seemed secondary. With the race over, the nation needed to scrutinize the bill and assess the gains to try to understand what had been accomplished, what should be sustained, and what should be discarded. The United States had gotten something for its money – great pictures, new understanding of the effects of spaceflight on the human body, and an appreciation of the overall difficulty of travelling in space. But few scientists could point to unique data which could not have been collected and analyzed at a substantially lower cost while posing substantially lower risk. NASA would spend the next two decades redefining its mission and justifying its work. This contrasted sharply with the gently rising NSF commitment to basic science which every year seemed to justify its cost. Real science might not produce economic or technological spin-offs, but it did increase general understanding of the universe. Skeptics of Apollo, both inside the scientific community and out, were not convinced that the space program had done the same.

Notes

1 NSF, "Position Paper: A National Effort in Space Exploration and Research," December 6, 1957, AW, 26: NSF, 1957, pp. 1–2.
2 Ibid., p. 6.
3 "National Space Program" (internal White House memo), OSAST 15: Space (3).
4 PSAC, "Introduction to Outer Space," 1958, RG 255, NASA/Vanguard, 2: Press Releases.
5 Richard Nixon, speech to the National Nuclear Energy Congress, March 19, 1958, RG 255, NASA/Vanguard, 2: Press Releases, p. 1.
6 H.W. Bode to Killian, February 24, 1958, DDE, PSAC, 4: NASA (3), p. 1.
7 Quoted in Enid Curtis Bok Schoettle, "The Establishment of NASA," in Sanford Lakoff, ed., *Knowledge and Power: Essays on Science and Government*, Free Press, 1966, p. 185.
8 "National Civilian Space Policy," January 12, 1961, JFK, WHCF, OS: 1961–62, p. 9.
9 Bode to Killian, February 24, 1958, p. 2.
10 Memorandum for J. R. Killian, February 21, 1958, DDE, PSAC, 4: NASA (3), p. 5.
11 National Space Program (internal White House memo), OSAST 15: Space (3).
12 Killian, Rockefeller, and Brundage, "Organization for Civil Space Programs," March 5, 1958, DDE, PSAC, 4: NASA (3).
13 Eisenhower Speech Before Congress, April 1, 1958, DDE, PSAC, 4: NASA (3).
14 Notably, the DoD had little opportunity to object to NASA's creation. Johnson had deliberately given Pentagon officials little time to comment on the bill, and the bill's authors had deliberately preserved nearly all military-related space programs within ARPA. Johnson would later comment that the draft "whizzed through the Pentagon on a motor cycle." Schoettle, "The Establishment of NASA," p. 238.
15 Audra Wolfe, *Competing with the Soviets*, Johns Hopkins University Press, 2013, p. 92.
16 NASA, "The Long Range Plan of the National Aeronautics and Space Administration," December 16, 1959, DDE, NASA, 1: OCB Working Group.

17 Lee Dubridge, "Policy and the Scientists," *Foreign Affairs*, 41, April 1963, p. 581.
18 NASA, "The Long Range Plan of the National Aeronautics and Space Administration," p. 16.
19 Both quoted in "Should the U.S. Enter Outer Space Race?" *Foreign Policy Bulletin*, December 15, 1959, p. 55.
20 Quoted in a memo to Alfred Loomis, "Material for Reply to Mer. Sherman Adams Concerning U.S. Moon Experiments," March 24, 1958, OSAST, 15: Space (4), p. 3.
21 Quoted in "The Space Race: How U.S. Became No. 2," United States *News and World Report*, October 19, 1959, p. 43.
22 King to JFK, April 15, 1961, JFK, WHCF, Outer Space: 1/18/61–1/25/62.
23 Welsh to JFK, February 14, 1964, LBJ, WHCF, OS, 1: 11/22/63–6/19/64. The Russell anecdote is reported in a routine memo from NASA's executive director to the White House. I could find no evidence for work toward death rays.
24 Kistiakowsky to McCone, March 16, 1960, OSAST, 11: H.E. Physics, p. 2.
25 PSAC, "To Race or Not to Race," October 16, 1959, OSAST, 15: Space (7), p. 3.
26 Ibid., p. 5.
27 Ibid., p. 2.
28 Ibid., p. 8.
29 "Statement by the President," March 26, 1958, RG 255, NASA/Vanguard, 2: Press Release.
30 See draft of letter from DDE to Khrushchev, 1958, DDE, PSAC, 4: NASA (3), p. 4.
31 Staff Report of the Select Committee on Astronautics and Space Exploration, *International Cooperation in the Exploration of Space*, GPO, 1958, p. 2.
32 "Attitude of Committee on Science and Astronautics Relative to the National Space Program," JFK, POF, 82: NSA 1/61–3/61.
33 CIA, "The Soviet Space Program," National Intelligence Estimate 11–1–65, LBJ, NSF – Subjects, 37: Soviet Space Shots, 2/2.
34 Also, NASA Memo, "Space Activities," October 16, 1964, LBJ, CF, 128: NASA 1967 1/2.
35 NASA Memo, "Space Activities," September 24, 1965, LBJ, WHCF, OS 2: 1/29/66–3/10/66.
36 From an interagency NASA memo, Scheer to Christian, Sherman, and Donnelley, October 18, 1967, LBJ, WHCF, OS 2: 7/19/67–8/17/67.
37 NASA Memo, "Space Activities," July 24, 1964, LBJ, WHCF, OS 1: 6/20/64.
38 Von Hardesty and Gene Eisman, "When Good Rockets Go Bad," in *Epic Rivalry: The Inside Story of the Soviet and American Space Race*, National Geographic, 2008, pp. 34–35.
39 Gus Grissom, Ed White, and Roger Chaffee died on January 27, 1967, in a fire in an Apollo command module in a simulated launch. (NASA later named the project Apollo I.) The atmosphere inside the capsule was almost pure oxygen making it impossible for the men to exit the capsule before being consumed by the flames.
40 JFK to Brooks, March 23, 1961, JFK, WHCF, OS:'61–'62.
41 Quoted in a JFK to Thomas, September 23, 1963, JFK, WHCF, OS: 1962–63, p. 1.
42 Ibid., p. 2.
43 Ibid., p. 3.
44 Anfuso to JFK, September 10, 1962, JFK, WHCF, OS: 61–62.
45 Quoted in Ford Eastman, "Kennedy Seeks U.S. Space Race Gains," *Aviation Week and Space Technology*, 74, October 9, 1963, p. 29.
46 From the *Washington Star*, October 7, 1957.
47 NASA, "Revised Draft of Guidelines for Public Information," RG 255, NASA/Vanguard, 2: Press Releases. NASA was not wholly unaware of Cold War considerations. It also directed its public affairs officers to avoid publicizing Soviet successes "gratuitously," and to "take advantage of opportunities to refer to the U.S. and other nations' programs."

48 "Questions from Everett Collier re President Johnson's Attitude on Space," LBJ, WHCF, OS 1: 11/22/63–6/19/64.
49 From the Vice President's Report on the Space Council," January 28, 1962, quoted in "Selected Quotations on Space of the President," LBJ, WHCF, OS, 1: 11/22/63–6/19/64, p. 1.
50 E.C. Welsh, June 18, 1964, quoted in "Selected Quotations on Space," LBJ, WHCF, OS, 1: 11/22/63–6/19/64, p. 1.
51 Ibid., p. 3.
52 Welsh to LBJ, January 14, 1964, LBJ, WHCF, OS 2: Aeronautical and Space Research, p. 4.
53 Ibid., p. 6.
54 "A Comparison of the American and Soviet Space Programs," LBJ, NSF 37: Soviet Space Flights, 1/2.
55 "Report of the Ad Hoc Panel on Man-in-Space," December 16, 1960, DDE, PSAC, 5: Space, pp. 1–4
56 NASA, "Report to the Congress," January 1962, JFK, POF, 82: NASA 1/61–3/61, pp. 3–8.
57 Data about the NASA manned missions are widely available. I took these statistics from "NASA Highlights," December 28, 1965, LBJ, CF, 128: NASA 1967.
58 White House, Press Release, February 9, 1967, LBJ, WHCF, OS, 2: 1/29/66–3/10/66.
59 NASA, "Report to Congress," 1964, LBJ, WHCF, OS, 1: 11/22/63–6/19/64, p. 2.
60 LBJ, "Statement," November 9, 1967, LBJ, WHCF, OS, 2: 7/19/67–8/17/67.
61 Special Space Review – 1965 budget, LBJ, NSF (subjects), 37: Space, Outer, Vol. 1, 2/3.
62 Donald Hornig, "Statement Before the Senate Committee on Aeronautical and Space Sciences," August 24, 1965, LBJ, DH, 7: Statements Before Congress 1965, p. 16.
63 "NASA Selects Apollo Guidance Contractor," August 9, 1961, JFK, WHCF, OS: 61–62.
64 Javits Cellar et al. to McNamara, August 29, 1962, JFK, WHCF, OS: 61–62. Also, Diehl to Donahue, February 14, 1963, JFK, WHCF, OS: 61–62.
65 O'Brien to Albert, November 20, 1962. Williams to JFK, January 22, 1963. Both in JFK, WHCF, OS: 62–63.
66 "Nixon Claims U.S. Dominates Space Race," *Aviation Week*, October 31, 1960, p. 29.
67 Webb to O'Donnell, April 21, 1961, JFK, WHCF, Outer Space: 1/18/61–1//25/62.
68 Statistics from Webb, "Memo for the President," August 9, 1963, JFK, POF, 82: NSA, 1/61–3/61.
69 Feldman to Welsh, undated, JFK, WHCF, OS: 1961–62.
70 Bush to Webb, April 11, 1963, JFK, POF, 82: NSA 1/61–3/61, p. 2.
71 Ibid., p. 3.
72 Ibid.
73 Bell to JFK, March?, 1962, JFK, POF, 82: NSA 1/61–3/61.
74 Wiesner to LBJ, April 9, 1963, JFK, WHCF, OS: L 1961–62.
75 Mary Lou Reitler to JFK, January 19, 1962, JFK, WHCF, OS: 1961–62.
76 Lendahl to JFK, April 9, 1962, JFK WHCF, OS: 1962.
77 Feldman to Reitler, March 29, 1962, JFK, WHCF, OS: 1961–62.
78 Hodges, Heller, Wiesner, "Technology and Economic Prosperity," December 3, 1962, JFK, POF, 85: OST 8/2/63, p. 2.
79 Ibid.
80 Quoted in Richard Austin Smith, "Now It's an Agonizing Re-appraisal of the Moon Race," *Fortune*, November 1963, p. 273.
81 Ibid., p. 7.
82 Smith, "Now It's an Agonizing Re-appraisal of the Moon Race," p. 128.

83 Ibid., p. 270.

84 Lee Dubridge, "Policy and the Scientists," *Foreign Affairs*, 41, April 1963, p. 582.

85 Webb to LBJ, January 14, 1964, LBJ, WHCF, OS, 2: Aeronautical and Space Research, p. 4.

86 Ibid., p. 9. In the author's opinion, the idea was reasonable. Boys, particularly, were drawn to, and inspired by, the Apollo mission. A program of visiting NASA scientists to the nation's public schools might have inspired a generation of potential scientists and engineers.

87 "News from Republican Citizens," May 28, 1964, OJB, WHCF, OS, 1: 11/22/63–6/19/64. Milton Eisenhower was President Dwight Eisenhower's brother.

88 Webb, "Memorandum for the President," July 10, 1964, LBJ, WHCF, OS, 1: 6/20/64.

89 From Donald Hornig, "Statement Before the Senate Committee on Aeronautical and Space Sciences," August 24, 1965, LBJ, DH, 7: Statements Before Congress, pp. 3–5.

90 Webb to LBJ, September 19, 1966, LBJ, CF, 74: Outer Space, p. 2.

91 Webb to Anderson, May 8, 1967, LBJ, CF, 128: NASA 1967 1/2, p. 2.

92 Stanford Research Institute, "Some Major Impacts of the National Space Program," LBJ, WHCF, OS, 2: 8/1/67–10/31/67.

93 Hornig, "Statement Before the Senate Committee on Aeronautical and Space Sciences," August 24, 1965, LBJ, DH, 7: Statements Before Congress, p. 4.

94 Ibid., p. 13.

95 Quoted in "New Leaders Will Overhaul United States Science Policy for 1970s," *Physics Today*, 22:2, February 1969, p. 73.

96 Webb to LBJ, July 11, 1968, LBJ, CF, NASA 1967, 128: NASA 1967 1/2.

97 Hornig, "Statement Before the Senate Committee on Aeronautical and Space Sciences," August 24, 1965, pp. 14–15.

98 "Space Goals After the Lunar Landing, October 1966, LBJ, NSF, 37: Space Goals After the Lunar Landing," p. ii.

12 Epilogue

After Apollo 11, the United States landed ten more astronauts on the moon over the next three and a half years. The final mission crew of Apollo 17 conducted geological tests, collected rock samples, drove a lunar rover, and carried five mice to the moon and back. It would be the last time that people stepped on the moon and the last use of the Saturn V assembly in its original intended usage.[1] NASA turned its attention to the Space Shuttle orbiter for manned spaceflight and to a series of unmanned missions which sent spacecraft to Mars, Jupiter, Saturn, Uranus, Neptune, and Venus. The Galileo spacecraft orbited Jupiter in 1995; the Sojourner landed on Mars and explored the surface in 1996; Genesis collected solar wind samples and returned them to the Earth in 2004; and Deep Impact collided with a comet (by design) in 2005. Many other spacecraft took pictures of Jovian moons, the sun, the moon, the asteroid belt, and Mercury. Our most recent mission, Perseverance, landed its rover on Mars in February 2021, and a return mission is planned for 2024.

The moon missions exemplified the tendency to invest in applied rather than basic science during the Cold War. Apollo was a triumph of engineering prowess, with the basic scientific components of the program almost an afterthought. After spending a staggering amount of money on very visibly putting an American on the moon, Congress would never again show such indulgence. NASA would need to justify its existence from the standpoint of scientific progress rather than mere feats of exploration, and missions henceforth would be evaluated on a stricter cost–benefit basis. A much-considered manned voyage to Mars has yet to fly or even be seriously planned. Each time it is proposed, its potential cost dwarfs possible scientific payoffs.[2]

But other patterns of federal scientific support laid down during the postwar decades continue. The NSF has grown to many times its original size (the most recent budget topped $8.5 billion) and continues to allocate its funds largely to university-based researchers who compete for grants through a series of peer-reviewed applications.[3] While fellowship support at the graduate and postgraduate level is now slightly more widely dispersed, a small group of research universities continues to glean the bulk of the grant

DOI: 10.4324/9781003363897-12

money. In recent years, nearly a third of the graduate students winning NSF support attended just ten schools and 14 percent attended just three.[4] Perhaps, more troubling, those students had attended only a select few undergraduate colleges: ten percent had attended Berkeley, MIT, and Cornell, and half of all award winners had attended just 30 schools. One scholar who studies the process noted that the fellowships were more "a feather in the cap, rather than a transformative resource."[5]

Federally funded research of all types continues to be similarly concentrated. Of the $75 billion allocated by federal agencies to all American colleges and universities for research in 2017, nearly 20 percent went to the top ten universities, and Johns Hopkins alone received $2.5 billion. The system has worked well for scientific output. The "tall steeples" envisioned by Frederick Terman in the 1950s have borne tremendous fruit. In the decades after Terman's reign, Stanford leaped to the very top ranking among the world's universities by many measures, with its endowment climbing above $20 billion and Silicon Valley becoming nearly synonymous with high-technology entrepreneurship. Other universities have tried to implement the model with greater or lesser success. MIT, Caltech, Berkeley, Harvard, Chicago, and Columbia have continued to invest in the highest quality faculty who can consistently garner substantial federal grants, as have scientists at the Universities of Wisconsin, Michigan, and Washington. Duke, Northwestern, and New York University have all made substantial gains following the Stanford model. In 2014, New York University announced its merger with Brooklyn Polytechnical University in an effort to create a stronger presence in the applied sciences, and Harvard, which had historically ceded engineering to its neighbor at the other end of Cambridge, announced the creation of its own engineering school in 2007. Within only a few years, that school was enrolling nearly a fifth of all Harvard undergraduates.

The Department of Energy, the successor agency to the AEC, continues to run its expanded series of national laboratories through contractual relationships with universities. It broke off weapons work from the Lawrence Berkeley Lab in 1971 when it created the Lawrence Livermore facility for that purpose, but it continues to run Los Alamos, Argonne, and Brookhaven with its original partners. Likewise, the Jet Propulsion Laboratory continues to partner with Caltech; the Applied Physics Lab is now a major component of the Johns Hopkins research program; and Lincoln Laboratory is a substantial presence within the MIT research enterprise. The hybrid model of the government-owned university-managed laboratory has proven to be enormously beneficial to American science and to those universities fortunate enough to become partner institutions.

By many measures, American science remains preeminent. Most rankings of world research universities award about half of the top 50 slots to American schools despite the fact that the American population constitutes only 4 percent of the world's total. Doctoral students and postdoctoral fellows from around the world flock to American research programs, hoping to attach themselves to the generous federal grants earned by their faculty mentors and gain access to the superb laboratories and equipment at the nation's top research schools.[6] Young scientists know that they enhance their professional prospects by starting their careers at the strong research schools even if they fail to stay; early career success enabled by simply being at these schools can lead to a lifetime of productive research.

But the concentration of talent impelled by federal research funding continues. The large, robust, and wealthy schools have become stronger and wealthier still, even as the majority of all American colleges struggle to attract capable students and hire competitive faculty. Just 10 percent of American universities (about 250) hold almost three quarters of all of the endowment funds held by universities, while the wealthiest ten universities hold over 40 percent of all the wealth. Not coincidentally, these wealthiest universities are also the most successful in attracting federal research funding. The two phenomena are related in complex ways, but history suggests that the federal funding came first. Those universities able to attract the first wave of significant federal research funding in the 1940s and 1950s were able to leverage these funds to attract the best faculty who brought with them the most competitive graduate students and later the most capable undergraduates. These students, in turn, were able to produce many of the greatest personal fortunes of their generation; a not insignificant portion of which found its way back to the universities in the form of gifts and bequests. The ten largest gifts ever to American universities, ranging from $350 million for Cornell to $1 billion to Johns Hopkins, nearly all went to schools that rank among the top ten in garnering federal research funding.[7] While it was never the intention of ONR, the NSF, or the AEC to enrich a handful of select schools, that has been one of the effects of federal grantmaking and contracting.

We are left questioning the wisdom of the system we have created. Like so many aspects of American life, we have created a "winner takes all" competition in university life, or at least a winner takes most. Competition for access to our elite universities has grown more intense year after year; in the midst of the global Coronavirus pandemic the strongest American universities are reporting record numbers of applicants, while the weakest struggle to keep their doors open. The wealthiest and best funded universities can now charge the highest prices, while most weaker schools must discount tuition heavily if they are to bring in a class. And the great state schools — California, Virginia, Michigan, Wisconsin, North Carolina — now seek to maintain fiscal parity by charging out-of-state tuition which approaches that of the most expensive private schools, with little financial aid on offer. As the most competitive students, often from highly functional and well-resourced families, concentrate themselves further in a small number of wealthy and well-funded colleges and universities, the influence of these schools grows. We have created a tiered system of higher education in which the great bulk of the nation's research and scholastic talent is concentrated on just a few campuses.

The system which the United States put in place under the leadership of Vannevar Bush, John Steelman, Alan Waterman, and Lee DuBridge has been tremendously beneficial to science, and arguably was critical in driving the Soviet Union into insolvency and political death. In a very real way American science won the Cold War, just as it had played a critical role in winning World War II. But perhaps we need to ask ourselves if the current extreme inequities in our system of higher education, which exacerbate and concretize inequities in American life, are necessary. Other nations maintain research institutes independent of their universities or spread government wealth more broadly throughout the university sector. Most nations fund their public universities at substantially higher levels than does the United States and flatten disparities between different colleges and universities. The United States is nearly unique in taking its most competitive young people, often from the most fortunate of backgrounds, and diverting substantially *more* educational resources toward them. Our science policy, conceived in the most frightening days of the Cold War, was not explicitly designed to do this but effectively has. It might be time to rethink our system of science funding and even our educational priorities.

Do we fund science generously enough? For the 75 percent of NSF grant applicants who submit proposals which are determined to be "competitive" but fail to gain funding, the answer is no.[8] Many gifted young scientists are driven from research careers by the inability to reliably obtain outside (usually government) funding for their work – a stipulated prerequisite to their continued employment on university faculties. Whether more funding would actually produce better science, however, is nearly unknowable. A substantial portion of all peer-reviewed research across all scientific fields will never be cited – an indication that the scientific community does not deem the work relevant, important, or promising.[9] We would not expect all scientific work, particularly not basic scientific research, to produce good results, however. The essential quality of basic research is its exploration of the unknown, and we should not be surprised that a significant portion of these explorations produce only marginal discoveries.

Applied research, by contrast, usually done on a contractual rather than grant-funded basis, produces more consistently useful outputs. The universities and private labs which bid on the contracts generally have long records of successfully completing their contractual obligations, and most would not sign contracts to produce data and goods which they did not realistically feel that they could deliver. While programs may run past deadline or over budget, they rarely fail catastrophically. Here it is easier to determine whether or not we are funding the programs adequately, as agencies doing the contracting are either achieving their stated goals or not. The nation's continued military supremacy and the successes of NASA over the past 40 years seem proof enough that our applied research program is functioning well. Whether or not more funding could have produced a stronger military, greater breakthroughs in fusion research, or more spectacular astronomical voyages are questions which go beyond the scope of this book. Those are questions more about national priorities than science policy.

However, the relationship forged between the federal government and private and state universities in the time period examined in this book has endured. The government–university axis, responding to scientific need for funding and freedom, has produced a structure uniquely successful in the history of science. As discussed at various points through this book, both basic and applied scientists require *freedom* to do their best work: freedom to work at their own pace, to explore questions of interest to them, and to design experimental protocols which they believe can answer those questions. In this sense, scientists are not unlike artists – uniquely creative and productive – but frequently unaccountable. In placing scientists working on behalf of the government within the protected sphere of universities or in university-managed complexes, we have insulated them from the normal pressures of commerce and bureaucracy while giving them the requisite support and funding critical to excellent research. Our nation was blessed with visionary scientific administrators who recognized that regardless of the urgency of the research enterprise, a good deal of the most innovative work would need to be outsourced to largely unsupervised university-based scientists. Allan Waterman recognized this in the substantial grant program he developed while serving as the director of research at the Office of Naval Research, and he continued this pattern when he took over as the founding director of the NSF. Commissioners of the AEC and members of the National Science Board recognized the critical role of independent university-based research to the nation's scientific enterprise and maintained the structure of the national laboratories through the Cold War and after. And the DoD continues, to this day, to allocate over half of its research budget to external university-based grants and contracts.[10] The structure, conceived amid wartime and Cold War pressures, has endured for 70 years. For American science, we have yet to produce a more productive relationship.

Notes

1 It would later be used in the Apollo–Soyuz docking mission and to ferry astronauts to the orbiting Skylab.
2 Both Presidents Bush proposed manned trips to Mars, with Bush 43 announcing a project in 2004, with the general goal of returning to the moon and exploring the planets beyond. Congress, however, was unsupportive.
3 The NSF has made efforts over time to dedicate specific grants programs to junior scientists, women scientists, and minority scientists, so the program is now more demographically diverse. But geographically and institutionally, it is as concentrated as ever.
4 The University of California at Berkeley, MIT, and Stanford.
5 Quoted in Jane Hu, "NSF Graduate Fellowships Disproportionately Go to Students at a Few Top Schools," *Science Magazine*, August 26, 2019, https://www.science.org/content/article/nsf-graduate-fellowships-disproportionately-go-students-few-top-schools.

6 For a firsthand account of foreign graduate students in the United States, see Irina Filonova and Paola Barriga, "Coming to America," *Inside Higher Education,* October 12, 2020, https://www.insidehighered.com/advice/2020/10/12/advice-international-graduate-students-who-come-to-study-us-opinion. Also, Lucas Laursen, "Coming to America: Doing a Postdoc in the US," *Science Magazine,* January 1, 2010, https://www.science.org/content/article/coming-america-doing-postdoc-us.

7 With some exceptions. Philip Knight has made numerous substantial gifts to the University of Oregon, and an anonymous donor gave $350 million to Rensselaer Polytechnic Institute in 2001. Allana Akhtar, "The Fifteen Biggest Private Donations to Universities by the Ultra Rich," *Business Insider,* May 21, 2019, https://www.businessinsider.com/the-biggest-private-donations-to-universities-by-individuals-2019-5.

8 www.nsf.gov/homepagefundingandsupport.jsp.

9 The figures for the portion of articles never cited is contentious. Dahlia Remler estimates that about 27 percent of articles in the natural sciences are never cited although admits that the percentage depends on the methodology used. See Remler, "Are 90% of Academic Papers Really Never Cited?," *LSE Blog,* April 23, 2014, https://blogs.lse.ac.uk/impactofsocialsciences/2014/04/23/academic-papers-citation-rates-remler/.

10 Congressional Research Service, "Defense Acquisitions: How and Where DoD Spends Its Contracting Dollars," July 2, 2018, CRS R44010.

Notable Persons in American Science Policy, 1945–1969

Baruch, Bernard An American financier who chaired the War Industries Board under Woodrow Wilson. He worked informally as an advisor to Franklin Roosevelt on wartime production during World War II. In 1946, he became the US representative to the United Nations AEC where he presented the Baruch Plan: an adaptation of the recommendations of the Acheson-Lilienthal Report calling for the international control of atomic weapons and uranium stockpiles.

Bronk, Detlev A biophysicist who was the president of the Rockefeller Institute for Medical Research (1953–1968) and of Johns Hopkins University (1949–1953). He was the president of the National Academy of Sciences from 1950 to 1962, served on the National Aeronautics and Space Council, and was an advisory member of the AEC.

Brooks, Harvey A physicist and dean of applied science at Harvard who served on the science advisory committees of Presidents Eisenhower, Kennedy, and Johnson. In addition to his position in the division of applied sciences, he held a secondary appointment at Harvard in the Littauer Institute for Public Policy.

Buckley, Oliver An electrical engineer who served as the president of Bell Labs (AT&T) from 1940 to 1951. He served on the General Advisory Committee of the AEC from 1948 to 1954 during which time he opposed the decision to develop the hydrogen bomb.

Bush, Vannevar A professor of electrical engineering and dean of engineering at MIT from 1932 to 1938. In 1938, he became the president of the Carnegie Institution of Washington, where he served until 1955 while also serving on the National Advisory Committee for Aeronautics (1938–1948), as the chairman of the National Defense Research Committee (1940–1941), and as the director of the Office of Scientific Research and Development (1941–1947). He was the author of *Science, the Endless Frontier* in 1945, which became the blueprint for federal science funding policy after World War II. He was also one of the principal creators and the first chairman of the RDB. In most assessments, he is considered the single most important figure in American science leadership during World War II and in formulating science policy during the immediate postwar years.

Compton, Karl A physicist and president of MIT from 1930 to 1948. He was also the president of the Society for the Promotion of Electrical Engineering (1938), a founder of the American Institute of physics, and the chair of Roosevelt's Science Advisory Board ((1933–1935). He was a member of the NDRC and a member of the Naval Research Advisory Committee from 1946 to 1948.

Compton, Arthur A winner of the Nobel Prize in physics who served as a professor of physics at the University of Chicago (1923–1945) and later as the president of Washington University of St. Louis (1946 to 1954). He authored an NDRC report on the potential for an atomic bomb in 1941. As the head of the Metallurgical Laboratory at the University of Chicago, he authorized the construction of nuclear pile #1 at Stagg Field, where Enrico Fermi created the world's first controlled, sustained fission reaction in 1942.

Conant, James Bryant A chemist and president of Harvard University from 1933 to 1953. He chaired the NDRC from 1941 to 1945 and served on the RDB after World War II. After resigning the Harvard presidency, he served as US High Commissioner for Germany.

DuBridge, Lee A physicist and director of the Radiation Lab at MIT during World War II. After the war he served as the president of the California Institute of Technology from 1946 to 1969 while also serving as science advisor to Presidents Truman and Eisenhower (1952–1955) and then to President Nixon (1969–1970). He served on numerous nonprofit boards and government advisory committees including RAND, the Carnegie Endowment for International Piece, the Institute of Defense Analysis, the National Science Board, and the Rockefeller Foundation.

Haworth, Leland A physicist who worked on radar at MIT's Radiation Laboratory, and after the war at Brookhaven National Laboratory, which he directed from 1949 to 1961. He served on the AEC from 1961 to 1963 and as the director of the NSF from 1963 to 1969. Upon retirement from the NSF he returned to Brookhaven, where he worked in senior administrative positions for the Associated Universities, the consortium of northeast universities which manage the lab.

Hornig, Donald A chemist who served as a group leader of the Manhattan Project at Los Alamos. After the war he taught at Brown University, where he was the dean of arts and sciences. He was recruited by President Kennedy to replace Jerome Wiesner as science advisor but did not start until January 1964, shortly after Kennedy's assassination. He served in that position until 1969, whereupon he became the president of Brown University.

Killian, James A career university administrator who served as executive assistant to Karl Compton at MIT and then held the presidency of MIT from 1948 to 1959. He served as the first Special Assistant for Science and Technology in the White House from 1957 to 1959, during which time he helped to establish NASA. He continued to serve MIT as the chairman of the MIT Corporation (the university's governing board) until 1971.

Kistiakowsky, George A physical chemist who served on the NDRC during World War II while working on the Manhattan Project. He later served as President Eisenhower's second Special Assistant for Science and Technology from 1959 to 1960. From 1962 to 1965, Kistiakowsky chaired the Committee on Science, Engineering and Public Policy at the National Academy of Science.

Lawrence, Ernest A Nobel Prize winning nuclear physicist at Berkeley who built the first cyclotron (a circular particle accelerator) in 1929. He built a series of progressively larger machines through the 1930s and created the basic structure of the lab, which would ultimately become the Lawrence Berkeley National Laboratory. After the war Lawrence built a 184-inch cyclotron largely using funds from the MED. He, along with Edward Teller, was instrumental in getting the AEC to establish a second nuclear weapons laboratory at Livermore, California in 1952.

Lilienthal, David A career utilities attorney and executive who served as the director of the Tennessee Valley Authority during the 1930s. In 1946, he coauthored (with Under Secretary of State Dean Acheson) the Acheson-Lilienthal Report, which recommended turning all nuclear weapons over to an international authority. When the Soviet Union vetoed the ensuing plan (the Baruch Plan) at the United Nations, Lilienthal became the founding chair of the AEC, serving until 1950. In the 1950s, he strongly advised against developing the hydrogen bomb.

Loomis, Alfred An investment banker and amateur scientist who established a private laboratory in Tuxedo Park, New York. He hosted eminent scientists from around the world and collaborated with E. O. Lawrence to construct a large cyclotron in 1939. He served on the microwave committee of the NDRC, and in 1940, transferred the bulk of his laboratory to Cambridge, Massachusetts, where it became the core of the new Radiation Laboratory at MIT. He is known as the creator of the Loomis Radio Navigation system, known today as LORAN.

Oppenheimer, J. Robert A physicist at the University of California, who served as the director of Los Alamos site of the Manhattan Project from 1942 to 1945. He directed the Institute for Advanced Study in Princeton, NJ after 1947 and served as the chairman of the General Advisory Committee of the AEC from 1947 to 1952. During the late 1940s, he strongly and vocally opposed the development of the hydrogen bomb.

In 1953, he was stripped of his government security clearance due to his association with known communists.

Piore, Emanuel A physicist, the director of research at the Office of Naval Research in the 1950s, and later the director of research at IBM.

Rabi, Isidor Isaac A Novel Prize winning physicist who spent most of his career at Columbia University. He worked at MIT's Radiation Laboratory during World War II and served on the General Advisory Committee of the AEC from 1952 to 1956. He also served as an informal science advisor to President Eisenhower and was active in the effort to establish the Brookhaven National Laboratory.

Seaborg, Glenn A Nobel Prize winning chemist at Berkeley who discovered plutonium along with at least eight other elements. He was the associate director of the Radiation Laboratory at Berkeley after the war and was the chancellor of the University of California at Berkeley from 1958 to 1961 while also serving on the General Advisory Committee of the AEC. He served on President Eisenhower's PSAC and authored a major report (the "Seaborg Report") in 1960, which recommended continued federal funding of science. From 1961 to 1971, he served as the chairman of the AEC.

Steelman, John Assistant to President Truman from 1946 to 1953 who took on the duties which would eventually devolve to the White House Chief of Staff. He was particularly focused on issues of science policy, serving as the chairman of the Scientific Research Board from 1946 to 1947. He was the principal author of the Steelman Report in 1947, which enlarged on recommendations from Bush's *Science, the Endless Frontier*, and notably recommended that 1 percent of GDP be spent on research.

Strauss, Lewis An investment banker who served two terms on the AEC (1946–1950 and 1953–1958), the second time as chairman. He was one of the principal proponents for stripping Oppenheimer of his security clearance. Later, when nominated by President Eisenhower to become Secretary of Commerce, he failed to be confirmed.

Teller, Edward A physicist best known as the "father of the hydrogen bomb," which he mused about even while working on the fission process during the Manhattan Project. He returned to Los Alamos in 1950 to work on a fusion weapon and developed a workable design while working with Stanislaw Ulam. In 1952, he left Los Alamos for the new Livermore Lab. Teller was a proponent of greater civilian use of atomic bombs (such as in Plowshare and Chariot) and attacked Oppenheimer for being overly pacifistic. In 1954, he testified that Oppenheimer was a security risk, and in so doing incurred the antipathy of much of the physics community. He directed the Lawrence Livermore Laboratory from 1958 to 1960.

Terman, Fredrick A professor of electrical engineering at Stanford, dean of engineering (1945–1955), and provost (1955–1965). During his time in administration he was a transformative leader, elevating Stanford engineering to the highest ranks of American engineering programs and turning the university toward a model of funded research in the engineering and natural science fields. He strongly encouraged the development of local high-technology firms near the Stanford campus and is often viewed (with William Shockley) as the cofounder of Silicon Valley.

Tuve, Merle A geophysicist who was the founding director of the Applied Physics Laboratory at Johns Hopkins University where the proximity fuse was developed during World War II. After the war, he also served as the director of the terrestrial magnetism laboratory at the Carnegie Institution from 1944 to 1966.

von Braun, Wernher A rocket engineer who was responsible for developing the German V-2 rocket during World War II. He was brought to the United States in 1946 with approximately 400 of his team members under Operation Paperclip, after which he was assigned to the Redstone Armory in Huntsville, Alabama. As the director of the Army Ballistic Missile Agency there, he directed the team that developed the Redstone and Jupiter rockets – the latter responsible for putting into orbit the Explorer I satellite in 1958. In 1960, von Braun's team was transferred to NASA's Marshall Space Flight Center (still in Huntsville). From 1960 to 1970, as the director of the Marshall facility, von Braun worked on the Mercury and Apollo programs and was instrumental in developing the Saturn rocket.

Waterman, Alan A professor of physics at Yale who served as the director of field operations for the OSRD during World War II, then as the director of research for the Office of Naval Research from 1946 to 1951. From 1951 to 1963, he served as the founding director of the National Science Foundation.

Webb, James Director of NASA from 1961 to 1968 who oversaw the Mercury, and Gemini programs, and laid much of the groundwork for the Apollo program.

Wiesner, Jerome An electrical engineer who worked at MIT's Radiation Laboratory during World War II and then at the successor Research Laboratory in Electronics from 1946 to 1961. In 1961, President Kennedy appointed him the White House science advisory and chair of the National Science Committee (1961–1964), during which time he was a vocal opponent of the manned space flight program. Wiesner returned to MIT after his White House service and served as the president of the university from 1971 to 1980.

Abbreviations Used in Notes

AW   **Paper of Alan Waterman (Library of Congress)**
GC   General Correspondence
NDRC   National Defense Research Committee
NSF   National Science Foundation
ONR   Office of Naval Research
OSRD   Office of Scientific Research and Development
PSAC   President's Science Advisory Committee

DB   **Papers of Detlev Bronk (Rockefeller Archives)**

DDE   **Dwight D. Eisenhower Presidential Library**
CCF   Clifford Cook Furnas Papers
DQ   Donald Quarles Papers
NASA   National Aeronautics and Space Administration
ODM   Office of Defense Mobilization
OSAST   Office of the Special Assistant for Science and Technology
PSAC   President's Science Advisory Committee
RC   Richard Cook Papers
WHCF   White House Central Files

FDR   **Franklin Delano Roosevelt Presidential Library**
POF   President's Official Files

GS   **Papers of Glenn Seaborg (Library of Congress)**

HST   **Harry S Truman Presidential Library**
PSRB   President's Science Research Board
SAC   Science Advisory Committee
SRF   Student Research Files
SRB   Records of the President's Scientific Research Board
PSF   President's Secretary's Files
WG   William Golden Papers

IIR	Papers of Isidor Isaac Rabi (Library of Congress)

JFK — **John F. Kennedy Presidential Library**
AE	Atomic Energy
JW	Jerome Wiesner Papers
NSA	National Security Agency
NSF	National Security Files
OS	Outer Space
OST	Office of Science and Technology
POF	President's Office Files
PSAC	President's Science Advisory Committee
PSF	President's Secretary's Files
RG	Roswell Gilpatric Papers
WHCF	White House Central Files
WK	William Kaufman Papers

LBJ — **Lyndon Baines Johnson Presidential Library**
AT	Atomic Energy
CF	Confidential Files
DH	Papers of Donald Hornig
FG	Federal Government
IO	International Organizations
OS	Outer Space
NSF	National Security Files
SC	Science
WHCF	White House Central Files

LH	Papers of Leland Haworth (Library of Congress)

MIT — **Massachusetts Institute of Technology, Office of the President**
NDRC	National Defense Research Committee
SAB	Science Advisory Board
SAC	Science Advisory Committee

NARA — **National Archives and Records Administration (Archives II)**
AEC	Atomic Energy Commission
NACA	National Advisory Committee for Aeronautics
NASA	National Aeronautics and Space Administration
NRC	National Research Council
OSRD	Office of Scientific Research and Development

VB	**Papers of Vannevar Bush (Library of Congress)**
NACA	National Advisory Committee for Aeronautics
NAS	National Academy of Science
NRC	National Research Council
OSRD	Office of Scientific Research and Development
RDB	Research and Development Board

Index

For Product Safety Concerns and Information please contact our EU
representative GPSR@taylorandfrancis.com
Taylor & Francis Verlag GmbH, Kaufingerstraße 24, 80331 München, Germany